The Evolution Delusion

By

Bo Kirkwood

truth
BOOKS

ISBN 10: 1-58427-407-7

ISBN 13: 978-1-58427-407-0

Guardian of Truth Foundation
CEI Bookstore
220 S. Marion St., Athens, AL 35611
1-855-492-6657
www.CEIbooks.com

Dedication
For Luke Michael, Carter, Kade, Micah, Clara, Noah, and Colette.

Table of Contents

Biographical Sketch

Dr. Bo Kirkwood is a Board Certified Family Physician in Pasadena, Texas where he has practiced for thirty-two years. He is on the faculty of two medical schools in Houston, Baylor College of Medicine and The University of Texas Health Science Center of Houston.

Before attending medical school, Dr. Kirkwood received a Bachelor of Science degree in Psychology and Biology from the University of Houston and a Master's degree in Behavioral Science from the University of Houston Clear Lake City. He then graduated from the University of North Texas Health Science Center before returning to his hometown to practice medicine.

Dr. Kirkwood is an elder at the Parkview Church of Christ and has lectured on evolution and creation for many years. His popular classes on these subjects can be found on YouTube by going to "Evolution Dr. Bo Kirkwood" or "Creation Dr. Bo Kirkwood."

Dr. Kirkwood is the author of *Unveiling the DaVinci Code* and co-authored with his two brothers, Dr. Ron Kirkwood and Dr. John Kirkwood, *A Case For Life: Christian Ethics & Medical Science*.

Preface

Evolution, the most pervasive scientific dogma of our time, is an absolute scientific fact, firmly established by extensive research and discovery, and those who do not accept it are either ignorant, mentally challenged or crazy, and further, those who actively oppose it are either religious zealots or have some secondary gain for doing so. Being a creationist is similar to believing in a flat earth. All scientific evidence points to a naturalistic causation for life and there is absolutely nothing scientific about creationism as virtually all scientists accept evolution. Theodosius Dobzhansky famously said, "Nothing in biology makes sense except in light of evolution." Ernst Mayr wrote, "No educated person any longer questions the validity of the so-called theory of evolution which we now know to be a simple fact." Finally, Richard Dawkins has said it even more emphatically: "You cannot be sane and well educated and disbelieve in evolution. The evidence is so strong that any sane, educated person has got to believe in evolution."

The problem with the above statements is they are simply not true. In fact, evolution is arguably the most deceptive scientific dogma of our age and the consequences to society have been onerous. As a result of the acceptance of the evolution paradigm, society has accepted relativity in morals and the concept of free will has been discarded. Furthermore, euthanasia and even eugenics now fit into the new ethic of our time.

The purpose of this book is to debunk the nonsense of evolution. Those reading this book will come to understand the molecules-to-man evolution theory cannot be established by the scientific method any more than can creationism and is more philosophic than scientific, with that philosophy being naturalism.

Since a young man I have always had a lot of interest in biology and science. In high school, like most public school students, I was taught evolution, which was especially emphasized in my biology II class my senior year. This class was a very good one taught by two enthusiastic teachers who had a definite passion for their work. In this class, when evolution was taught, the teachers, in particular one, Mr. Collins, requested that those who did not believe in evolution defend their stand. From his standpoint, evolution was a fact. From things I had been taught, I was able to give my critique, but his response to that critique indicated to me he had never heard some of my arguments. Interestingly, in that same class, experiments with the common fruit fly, *Drosophila melanogaster*, seemed to disprove at least one tenet of Neo-Darwinism. In genetic experiments with these flies, we induced mutations and some of the mutations were rather dramatic, but in every case these mutations were detrimental to the organism.

As a pre-med student in college I was exposed to evolution in many courses, including Comparative Anatomy and Physiology. As an elective, I decided to take a course called Population Genetics and Evolution, as I was interested in learning more about evolutionary thought. To my surprise, the first day of the class at the University of Houston, the professor emphasized that his course was about microevolution (adaptation) and genetics, and stated the general theory of evolution belonged in a philosophy class and not a science class. This was the first time I had personally heard a scientist of this caliber say something like that.

I began medical school in 1978, and though evolution was not taught specifically it was understood, especially in classes such as microbiology and anatomy. During a discussion with a classmate, he found it astonishing that I did not believe in evolution, or at least molecules-to-man evolution, and when I pointed out the fossil records did not support evolutionary theory, it was apparent he had never heard these ideas before. This was not really a reflection on my friend, as most medical students and hence most doctors have never taken time to study evolution; they simply accept it.

Sadly, I believe my classmate's ignorance of the arguments against evolution are commonplace amongst scientists who perfunctorily accept it, not because it is a proven theory but simply because it is naturalistic. They never involve themselves with a detailed study of the hard evidence; it has no priority for them. The problem is in a person's world view. Most scientists refuse to consider a spiritual view of the universe; everything must therefore be explained from a purely materialistic, naturalistic causality. To consider otherwise is not an option.

In a *Houston Chronicle* article in November of 2013, entitled "It's Time for Science Texts to Evolve," two professors of biology, one from Rice

University, the other from Southern Methodist University, wrote an editorial in response to a proposed request that textbooks in the state of Texas include the mention of intelligent design when discussing the origins of species. These authors wanted to send a message that "Sound science was at stake." They accused creationists of being ideologues and dishonest. They stated, "Transitional fossils occurred in the thousands in nature," but offered no examples. They further alluded to the Scopes trial as proof for evolution which, of course, it is not! They wrote that evidence for evolution was conclusive and non-debatable, and to include intelligent design in textbooks would undermine the education of our youth and even possibly dissuade scientists from coming to Texas.

The dishonesty, it seems to me, lies with the evolutionists, not the creationists. Textbooks need not "evolve" because they currently contain nothing of the intelligent design arguments. Why is the mere mention of intelligent design so abhorred by the scientists? Who is more open-minded in this regard?

In this book you will see convincing evidence that evolution, from a logical, statistical point of view, is virtually impossible. One has to discard common sense completely to believe in molecules-to-man evolution.

You might ask, "Is it important for someone to know the facts as they pertain to evolution?" I believe it is. Making no apologies, my ultimate purpose in writing this book is to "convict the gainsayer," opening the possibility of a creator and eventually bringing that person to Christ, and helping others do so by fairly looking at the facts. Creationism gets no air time at all in our classrooms today. You will not get the information contained in this book in a public classroom.

But can't a person believe in evolution and still be a faithful Christian? Theoretically that might be possible, but practically it is very doubtful. Evolution undermines the creation story of the Bible and makes atheism much more palatable. Many have turned to atheism after accepting evolution. Even Richard Dawkins claims to have at one time been a believer but once he accepted evolution, atheism was obvious. Unfortunately, his story can be retold many, many times.

Evolution must be reduced to the so-called theory that it is, although it has never undergone the rigors of the scientific investigation required to be a theory. It is my hope that after a fair, open-minded reading of this book, you will see that evolution takes more faith to accept than creationism.

Acknowledgements

This book is impossible without the help of others which I am so indebted to. Much of this book came from a series of lectures on evolution and creation. For his efforts in helping in the organization and production of those classes I thank Norris Long, who originally suggested the use of mass media in the presentation. His insistence on detail and perfection were truly appreciated. I would also like to thank all the students of these lectures as well as students of past years who have sat in my classes, offering important feedback and making teaching a rewarding and pleasurable experience.

I would also like to thank April Mitchell in the initial transcription of the manuscript and Lory Guajardo in her administrative efforts in getting the book to print. Further, I would like to acknowledge Sherry Cecil for proofreading this work and for her suggestions and corrections.

Also, I would like to express an extreme amount of gratitude for those who critically reviewed my book, Dr. Joshua Gurtler, Dr. Craig Thomas, and Rick Billberry. Their recommendations and suggestions were very helpful and improved the acumen of this work significantly. I thank them so much for all their effort and input in making this a better book.

I am especially grateful to Mike Willis of Truth Books for agreeing to publish the book and for his expert critique and suggestions as well as the hours of work he put in getting it to its final form. Of course, I'm indebted to all those scientists who have publically acknowledged their belief in intelligent design and supernatural creation, many who appear in the chapters of this book. I would also like to thank Jerry Fite for his many years of friendship and encouragement, but also for his unyielding stand for the truth even in times when the truth is not so popular.

Finally, and most importantly, I would like to thank my beautiful wife of 42 years, Cherry. This work would have not occurred without her. Her noncomplaining attitude helped immensely in the many corrections and changes in the manuscript, plus the slide productions for my class, all requiring untold hours of work. Putting up with my own variations in attitude required supreme forbearance on her part and I am so happy and blessed to have her by my side. Much credit for this book goes to her. Thank you Cherry so much and I can not think of my life without you.

Bo Kirkwood
September 2015

Chapter 1

Introduction

What is truth? This question was sarcastically posed over 2,000 years ago by Pontius Pilate. There are those who believe we cannot know the truth. Post-modernists hold that subjectivism determines truth or reality for an individual. Some would say that the truth is, you cannot know the truth, which in itself is a contradictory statement. They would further argue that what is true for you is not necessarily true for me.

Jesus, on the other hand, said, "You shall know the truth and the truth shall set you free." Contrary to the post-modernist, truth is absolute and not relative. In fact, to believe relativism requires a belief in the truth of relativism, which is an "infinite regress."

Think about this: the physical sciences would be impossible to study if there were no absolute truths such as gravity, electromagnetism, and the speed of light. Ultimately, the purpose of science is to uncover or discover these fundamental truths. To discover truths, scientists postulate theories. Stephen Hawking, in his book, *A Briefer History of Time*, writes, "A theory is a good theory if it satisfies two requirements. It must accurately describe a large class of observations on the basis of a model that contains only a few arbitrary elements, and it must make definite predictions about the results of future observations." Furthermore, a good theory makes predictions that can be disproved, or falsified, by observation and experimentation.

Good science should be guided by a sincere search for the truth and not biased based on one's core philosophical beliefs. This, however, can be very difficult to achieve. Naturalism is the prevailing philosophy of most scientists today and is the reason many, but certainly not all of them, disregard creationism. They do so not because naturalism is more scientific, because it isn't. Good science is practiced, and has been practiced in the past, by scientists who espouse a creationist paradigm, which makes Dobzhansky's assertion that "nothing in biology makes sense except in light of evolution" an absurd statement. Isaac Newton, Johannes Kepler, and John Clerk Maxwell, to mention a few scientific geniuses of the past, were all creationists, as are many more recent scientists.

This brings us to the question: What is science? Dr. Danny Faulkner, astrophysicist, has written, "Science is the study of the natural world using the five senses." Francis Bacon, centuries past, described science as posing a question or hypothesis. You never prove a hypothesis, but more generally, develop experiments that try to contradict or disprove it. When experimentation cannot disprove the hypothesis, then that theory becomes much more relevant. As Faulkner has put it, "A theory is a grown up, well developed hypothesis." This whole process is what constitutes the so-called scientific method.

Historical sciences cannot really be true science because of the inherent difficulty in falsification as it is described. As a result, cosmologists who study the origins of the universe observe the natural phenomena and the cosmos for facts or evidences that provide the basis for a specific theory or viewpoint. Facts are facts (if they be facts); it is how one interprets those facts, and one's biases always come into play, which of course, should not be the case in true natural sciences.

Man postulates theories and then tries to prove or refute them. This frequently requires us to abandon previous postulates. The astute know that we don't know everything; man still does not have a complete understanding of the physical universe and its laws. Richard Feynman, the brilliant physicist of the 20th century, has written, "Each piece or part of the whole of nature is always merely an approximation to the complete truth or complete truth, so far as we know it. In fact, everything we know is only some kind of approximation, because we know that we do not know all the laws yet. Therefore, they must be learned only to be unlearned."

Still it is hard for scientists to give up on strongly held theories. For example, at one time the prevailing thought was that the earth was the center of the solar system, the so-called Geocentric or Ptolemaic Theory, even though there were some puzzling phenomena astronomers could not explain. The theory was held as self-evident and true, even sacred, until Copernicus and later Galileo proved that the sun was the center of the solar system – and, as I am sure you are aware, they were met with great resistance until scientists could no longer ignore the facts. Similarly, scientists at one time believed that the cosmos and the heavenly bodies in the universe existed in an unseen substance known as the ether (also spelled aether). This was postulated as the medium for the propulsion of light. In the late 19th century, scientists at Case Western in Cleveland, Albert Michelson and Edward Morley, created an experiment that refuted the ether theory, then Albert Einstein with his special theory of relativity put the ether postulate to rest for good.

Other examples could be cited, but the point is scientists sometimes have a hard time giving up on theories they accepted for many years. Thomas Kuhn has coined the phrase, "The Priority of the Paradigm" to explain this phenomenon of scientists accepting as dogma a theory later proven incorrect. Michael Denton has said that scientists "will go to extraordinary lengths ... to defend a theory just as long as it holds scientific intrinsic appeal." Referring to phlogiston, a widely held belief in the 18th century of a substance lost in combustion, now completely debunked, Professor Herbert Butterfield of Cambridge, quoted in Michael Denton's *Evolution: A Theory in Crisis*, comments ...

> The last two decades of the 18th century give one of the most spectacular proofs in history of the fact that able men who had the truth under their very noses, and possessed all the ingredients for the solution of the problem—the very men who had actually made the strategic discoveries—were incapacitated by the Phlogiston Theory from realizing the implications of their own work.

Paradigms are hard to change, and the priority of the paradigm can take precedence over logic and common sense sometimes. I have seen this in the medical field during my thirty years as a family physician. It was once taboo to use beta blockers in a congestive heart failure patient, and now it is mandatory. For many years, stress was felt to be the most common cause of gastric ulcers, until in 1982 two ingenious doctors in Perth, Australia, Robin Warren and Barry Marshall, proved bacteria were actually responsible for ulcers. Their work was met with intense criticism and widely doubted, until Dr. Marshall himself drank a bottle of the bacteria, now called *Helicobacter pylori*, which gave him an ulcer and then proved the ulcer was caused by H.Pylori by treating himself with an antibiotic that cured the ulcer. In 1994, twelve years after their initial discovery, the National Institute of Health in the United States finally recognized their work and accepted this new cause, and later new treatment, for peptic ulcers.

The great fallacy of our time is that evolution is scientific and creationism is not, even though the former is considered scientific dogma. The fact is, neither is scientific and both require a degree of faith. A scientist can no more prove evolution than a creationist can prove God. It is impossible to falsify past events by experimentation. This is not to say research is not being done in the name of evolution. On the contrary, billions of dollars, primarily provided by governmental sources, have been spent in studies either directly or indirectly related to the science of macro-evolution. How much is spent on creationism? Of course, that is a silly question since it is obvious that very little by comparison has been allocated by governmental sources for the study of creationism. What must be considered, however, is that evolutionary studies already presuppose evolution. No scientific studies are being conducted or have been conducted to prove or disprove evolution as it is already assumed. But there have, none the less, been numerous failed experiments by neo-darwinist documented in Jerry Fodor's book, *What Darwin Got Wrong.*

To determine which view is more correct, one has to look at the evidence. After investigating the evidence with as much of an open mind as one can, then and only then is it fair to decide which belief best fits the facts.

The problem is that many will never accept creationism solely because it is not naturalistic; they will either never lower themselves to study the evidence for creation, or they will simply perfunctorily ignore it. The truth is most scientists outside the biological sciences will never examine the facts. It is much more comfortable, and for that matter much less risky, to accept evolution than not. One is not allowed to question the evolutionary paradigm in the scientific world. Ben Stein's excellent movie, *Expelled*, chronicles the cases of several scientists whose careers were sabotaged for either teaching or even hinting about intelligent design. David Berlinski referred to this as "intellectual terrorism." Certainly, the media sides with the establishment, and pity the scientist, celebrity, or politician who publicly admits to a belief in creationism. He will certainly be subjected to ridicule and scorn for this belief.

The desire I have in researching and writing this book is to seek the truth, looking closely at the claims evolutionists make and then examining evidence that seems to contradict those claims. I make no apologies for my Christianity but my Christianity is a belief based on facts, not on some "better felt than told" subjective feeling. Being a Christian is logical and based on the interpretation of facts. I believe as Clark Pinnock has written, "The facts backing the Christian claim are not a special kind of religious fact. They are cognitive, informational facts upon which all historical, legal, and ordinary decisions are based."

How does this relate to evolution? The universe shouts of a creator. The apostle Paul, in Romans 1:20, stated, "For since the creation of the world his invisible attributes – his eternal power and divine nature – have been clearly seen, being understood through what has been made, so they are without excuse." The Psalmist, David, further spoke of evidences in Psalm 8: 1 when he wrote, "Oh Lord, our Lord, how majestic is thy name in all of Earth, who hast displayed thy splendor above the heavens."

The physical universe, created by "laws" which I believe God instituted, reveals remarkable things, some of which even defy common sense, yet nonetheless give overwhelming proof of a creator of divine, infinite intelligence. Naturalistic processes can simply never explain "the creation of the universe." British astrophysicist Paul Davies, once an ardent atheist, conceded that the laws of physics "seem themselves to be a product of exceedingly ingenious design." He further has written, "There is for me powerful evidence that there is something going on behind it all. It seems as though somebody has fine-tuned nature's numbers to make the universe. The impression of design is overwhelming."

In the second part of this book we will examine more closely these physical laws, and the obvious design inherent in them that Davies alludes to. But when one accepts the likelihood of intelligent design and creation, it becomes more and more obvious that evolution can simply not work because the science does not support it. This being said, one can be difficult to persuade the skeptic of creationism by going simply to the Bible since that individual has not accepted a creator to begin with, although many have come to the truth by examining the claim of the Bible, especially as it pertains to its scientific, geographic, historical and prophetic accuracy. This becomes a *non-sequitor*. Hence, we look at the evidences, and that is what this work is all about.

I will not eliminate the Bible from my arguments with evolutionists, for to do so is to begin the argument on their terms. Paul states in the Roman letter, "For I am not ashamed of the gospel for it is the power of God for salvation to everyone who believes." I cannot divorce myself from the Word when discussing evidences with non-Christians, and doing so would give me no alternate interpretation of the facts anyway.

Most evolutionists propose that there is no God. Christians propose there is a God, and believe the Bible is His word. So when debating the facts, we must keep in mind we are arguing about interpretation of those facts. The evolutionists will have a significant problem realizing this because, as Ken Ham has put it, "A non-Christian can't put on the Christian's glasses." On the other hand, I believe most Christians can usually see the evolutionist's point of view, as they have been force-fed this in school anyway.

Do not be deceived. The non-Christian is not neutral. He begins with a belief that there is no God, yet the Bible is clear, "The one who is not with me is against me and the one who does not gather with me scatters" (Matt. 12:30). Agreeing to debate a non-Christian by leaving the Bible account of creation out will start the discussion on an un-level playing field.

In discussing the scientific evidence for creation, I will strive to avoid what I call "flaky evidence." Some well-meaning Christians in the past have searched for a "magic bullet" to prove creationism in the biblical account and have fallen prey, in some cases, to hoaxes, and in other instances questionable data. This serves no good purpose and can be counterproductive.

My approach here is to acknowledge my Christianity and belief in the Bible but also realizing science does not contradict it. The Bible and science go hand in hand when presuppositions are discarded or at least acknowledged. The Christian should not fear science; in fact, it is his ally, as this book I hope will bear out. That is not to say I have all the answers. This is not a copout as physicists are still searching for many answers, too, especially in trying to unify relativity and quantum mechanics. In fact, it may be that this side of heaven, mankind will never know everything but, not knowing everything doesn't mean we cannot know some things. With that we should keep in mind Proverbs 1:7, "The fear of the Lord is the beginning of knowledge."

This work, then, is a defense. It is a defense for creationism and a defense for the power and glory of God in His gospel. The apostle Peter has written in his first letter, chapter 3:15, "But sanctify the Lord God in your heart and always be ready to give a defense to everyone who asks you a reason for the hope that is in you, with meekness and fear." If one cannot defend his or her belief based on facts, then he or she is really no better than anyone whose belief is based simply on how he feels or what he was taught and simply blindly believes.

The first section in this book, then, will look at evolution. We will begin by examining the historical account and how Darwin developed his theory. We will then turn to the various evidences for the theory of evolution, ending with a discussion on why the theory continues to be ensconced in scientific dogma. In the second section we will turn to creation. There we will examine the various cosmologies and how they relate to the biblical account. Specifically, we will examine how fine-tuned the universe is, and the probabilities of chance alone explaining our origins.

This will not be an exhaustive discussion on every topic covered. Books have been written on each subject discussed, and where applicable, reference is made to these works.

Chapter 2

A Historical View: Origin of Species

In 1859, Darwin published *On the Origin of Species or The Preservation of Favored Races in the Struggle for Life*, which has been called "One of the most important books ever written." It sold out the first day of its publication and changed the course of history, and changed the interpretation of the world in its prevailing explanation amongst scientists for origins of life. For many, it is no longer a theory, but fact. Julius Huxley, in 1959, stated, "The first point to make about Darwin's theory, is that it is no longer a theory but a fact. ... Darwinism has come of age so to speak. We do no longer have to bother about establishing the fact of evolution." More recently, Richard Dawkins has written, "The theory is about as much in doubt as the earth goes around the sun."

Before *On the Origin of Species*, species were held to breed true to their type, generation after generation, according to the so-called typological model. Nineteenth-century comparative anatomists like Cuvier, Owen, and Linnaeus divided the living world into phyla based on this. Species were fixed, or "immutable," and this concept traces back to the time of Aristotle. Discoveries in science, particularly those of Isaac Newton, were taken as evidences of a creator or a grand designer. William Paley's famous book on evidences, *Natural Theology*, illustrated the widespread appeal of this. In 1857, two years before the publication of Darwin's *Origins*, Louis Agassiz, Professor of Zoology at Harvard, wrote, the living world "shows also premeditation, wisdom, greatness, prescience, omniscience, providence ... all these facts ... proclaim aloud the One God whom man may know, and natural history must, in good time become the analysis of the thoughts of the Creator of the Universe, as manifested in the animal and vegetable kingdoms, as well as in the inorganic world."

For many years before Darwin, catastrophism was the prevailing thought in describing geologic formations. Catastrophism proposed that great catastrophes explained all kinds of geologic formations. This view was widely held and accepted by most scientists. The above ideas, that is the typological model, and catastrophism, were not universally accepted, however. The possibility of transmutation had been considered by French biologists, Buffon and Jean-Baptiste Lamarck, and even Charles Darwin's grandfather, Erasmus Darwin. Charles Lyell in 1830 published his *Principles of Geology* where he introduced the concept of uniformitarianism and he, along with James Hutton, drove the change from catastrophism. This publication proved to be a watershed change in geologic thought. Lyell proposed that all agencies which produced geologic changes in the past were the same as we observe today. Specifically, Lyell denied the global flood mentioned in the book of Genesis. His denial of the flood is in stark contradiction to what the Bible states in 2 Peter 3:3, "Know this first of all, that in the last days mockers will come with their mocking, following after their own lusts, and saying, 'Where is the promise of his coming? For ever since the fathers fell asleep, all continues just as it was from the beginning of creation.' For when they maintain this, it escapes their notice that by the word of God, the heavens existed long ago and the earth was formed out of water and by water, through which the world at that time was destroyed, being flooded with water." Charles Darwin was greatly influenced by Lyell and in 1875, at the time of Lyell's death, Darwin wrote, "I have never forgot that almost everything which I have done in science I owe to the study of his great works."

Charles Darwin was born in 1809, February 12th, the same day as Abraham Lincoln, and lived during the middle of Victorian England, at a time when the sun never set on the British Empire. It was a time following the Age of Enlightenment, a time of Romanticism but still a time of tremendous changes in industry, politics, culture, and science. It was also a period when literacy increased and the population grew significantly, not only of England but the entire world.

Charles Darwin
(1809-1882)

Of aristocratic birth, Darwin was destined to become a physician, being both the son and grandson of doctors. Originally enrolled at Edinburgh to study medicine, young Darwin seemed to be more interested in shooting and sports than academics, and eventually transferred to Cambridge to study theology. Of course Mr. Darwin was never really interested in theology, but for the aristocracy of his time, theology seemed to be a secondary or tertiary afterthought. As Gertrud Himmelforb, a Darwin biographer has written, "The final recourse of Victorian society for the maintenance of misfits and dullards was the church. Young men with no discernible calling were graced with the highest calling of all."

Darwin, however, was no dullard, and while at Cambridge, he found his true calling, the natural sciences. Under the tutelage of Professor John Heslop, a botanist, and Professor Adam Sedgwick, a geologist, he began a course of study that would eventually lead him to his world travels and the development of a theory that would forever change the face of biology and even the world.

After graduating Cambridge, Darwin received an invitation to travel on the *Beagle* as a naturalist under Captain Robert Fitzroy. The expedition was to sail around the continent of South America and eventually to the Galapagos Islands, some 600 miles off the coast of what is now Ecuador. This proved to be the turning point of Darwin's life. During this five-year voyage, after observing birds, countless animals and plants, as well as collecting innumerable specimens, including those from the Galapagos, Darwin formulated his theory which he solidified after his return to England. This voyage of the *Beagle* became "symbolic of the much greater voyage which the whole of our culture subsequently made from the narrow fundamentalism of the Victorian era to the skepticism and uncertainty of the 20th century. Darwin's experiences during those five years became the experiences of the world," opines Michael Denton in his book, *Evolution: A Theory in Crisis*. Several years after finishing his voyage on the *Beagle*, Darwin began work on *On the Origin of Species* and completed it after much procrastination in 1859. For the rest of his life, Darwin would defend the book, although he was never completely dogmatic about his theory.

The theory of evolution did not just pop into Darwin's head during his voyage on the *Beagle*: the seeds of evolution had been planted long before. In fact, his grandfather, Erasmus Darwin, proposed that "all warm blooded animals have arisen from one filament." A contemporary of Charles Darwin, Russell Wallace, a British naturalist, wrote in 1855, four years before the publication of *On the Origin of the Species*, "Every species has come into existence coincident both in space and time with a pre-existing closely allied species." Wallace felt new species arrive by the progression and continued divergence of varieties that outlive the parent species in the struggle for existence, a belief Darwin knew about and incorporated in *Origins*, claiming he and Wallace arrived at this conclusion at the same time.

The final piece of the puzzle for Darwin came from Thomas Malthus, a political economist of the late 18th and early 19th centuries. For Malthus misery was built into the human experience. He believed poverty and famine were natural outcomes of population growth and a diminishing food supply. He felt this was God's way of punishing mankind for laziness. Years after writing *Origins*, Darwin in 1876 wrote in his autobiography,

> In October 1838, that is, fifteen months after I had begun my systematic inquiry, I happened to read for amusement Malthus on *Population*, and being well prepared to appreciate the struggle for existence which every-where goes on from long-continued observation of the habits of plants and animals, it at once struck me that under these circumstances favor-able variations would tend to be preserved, and unfavorable ones to be destroyed. The results of this would be the formation of a new species. Here, then I had at last got a theory by which to work with.

Population growth and over-production for Darwin was a means of survival for the species.

Darwin postulated his hypothesis on the observation that different species seem better adapted for the habitat than others. Although a lot has been written about Darwin's finches of the Galapagos, they are never mentioned in *Origins*. Instead, much of Darwin's theory was based on observation of domestic animals and breeding practices he observed in his native England. A good deal of attention is given to pigeons, dogs, horses, cattle, and other domestic animals. What Darwin saw was a tremendous variability in these various species but he understood and wrote in *Origins*, "… the mere existence of individual variability and of some few well-marked varieties, though necessary as the foundation for the work, helps us but little in understanding how species arise in nature."

What Darwin proposed was that a single species diverges into several varieties, then, into several different species, through the process of natural selection, which was really the brilliance of his theory. Darwin believed this occurred over time when, through environmental factors, one variety would have a certain advantage over another to the point where the disadvantaged would eventually become extinct. The change into a different species would, of course, require "transitional forms" and Darwin knew that these forms would be "innumerable." Obviously, this change from one species to another would take vast amounts of time, another fact Darwin acknowledged in *Origins*.

As expected, Darwin's book created a hail storm of controversy. Captain Fitzroy of the *Beagle* became obsessed with his self-imposed blame for the anti-Christian influence of *The Origin*. Fitzroy was a devout believer and had given Darwin Lyle's book on uniformitarism that so influenced his thinking. As a consequence, Captain Fitzroy eventually committed suicide. Darwin's book created a change in how mankind now viewed the world and has influenced scientific thought ever since.

For Darwin, gradualism and immense amounts of time were "fundamental pillars" to his theory. In fact, it has been said that geologic uniformitarianism eased the way for acceptance of Darwin's theory. Darwin's theory was relatively simplistic. Natural selection, later called survival of the fittest (by Herbert Spencer), through trial and error, resulted in the formation of life and its many variations that followed. Darwin proposed, as stated by Jacques Monod, "… That chance alone is at the source of every innovation, of all creation in the biosphere. Purely chance, absolutely free but blind is at the very root of this stupendous edifice of evolution." In essence, life then is a giant lottery!

In *On the Origin of Species*, Darwin presented really two different theories, one referred to as the "special theory" and the other as the "general theory." The special theory proposes that new races and species arise in nature by the agency of natural selection. The more radical "general theory" applies the special theory universally, and purports that the appearance of all life came from a common beginning and can be explained by the same processes that caused trivial changes observed on the Galapagos Islands and elsewhere. As stated by Michael Denton, "If the *Origin* had dealt only with the evolution of new species it would never have had its revolutionary impact."

For Darwin, homology, that is the resemblances in structure, was highly suggestive of the reality of macroevolution. Darwin pointed to the existence of "rudimentary organs" as further proof of his theory. Rudimentary organs – later referred to as vestigial organs – were the "left-overs" of an organism, that once served a vital purpose but not any longer. Also, what Darwin saw as a succession of related types over periods of millions of years seemingly gave proof to his theory.

Though Darwin never claimed his theory could explain the origin of life, his book would tend to suggest otherwise. Observe this passage from *Origin*, "It is often said that all conditions for the first production of a living organism are present, which could ever have been present. But if (and Oh! what a big if!) we could conceive in some warm pond, with all sorts of ammonia and phosphoric salts, light, heat, electricity, etc., present, that a protein compound was chemically formed ready to undergo still more complex changes, at the present day such matter would be instantly devoured or absorbed, which would not have been the case before living creatures were formed."

Darwin proposed no mechanism for the change in species and wrote, "the laws governing inheritance are quite unknown," even though he was a contemporary of Gregor Mendel, the "father of modern genetics." In fact, Darwin proposed a Lamarckian-like inheritance mechanism called pan-genesis. In his concept, acquired characteristics were passed down through germ (sex) cells to their progeny.

Mendel, a somewhat obscure Austrian monk/scientist, working with peas, disproved the concept of blending and the Lamarkian concept of acquired inheritance, and his experiments actually put limits on organic variability and hence limits on natural selection. To be fair to Darwin, Mendel's work was not widely known of at that time, but had it been, one does wonder how it would have changed evolutionary theory.

Of course, it would be 100 years before the structure of DNA would be discovered by Watson and Crick and then evolutionist had their mechanism: mutations. Of course scientists knew of mutations before but they had no defined mechanism on how they could cause evolution to occur. Through mutations or changes in the sequencing of the DNA "code," species could change, or so the "Neo-Darwin" theory proposes. For the neo-Darwinist microevolution produces macroevolution through mutations and natural selection. This will be discussed in more detail later, but even though 19th century scientists knew nothing about how evolution could occur, the theory was readily accepted by much, if not most, of the intellectual community.

Darwin was keenly aware of the biggest obstacle with his theory and that was the lack in the fossil record of transitional forms which we have seen should be "innumerable." Darwin gives several reasons for this lack of evidence for his theory, but primarily he felt it was the "record being incomparably less perfect than is generally supposed," but even he knew this was not satisfactory. He felt in time the record would supply the necessary transitional forms that would validate evolution.

In retrospect, it seems somewhat amazing that Darwin's theory was so readily accepted during his time and it is fair to say the way was paved by the Enlightenment period of the previous two centuries. During this "Age of Reason," philosophers began to challenge institutions rooted in tradition such as the church, emphasizing free speech and free thought. Reason over superstition became the theme during this era, and it was at this time that naturalism had its beginning and man began looking towards science more and more. Also, it was during this time that the so-called scientific method was developed.

When evolution was introduced it would seem the intelligentsias of the day were ready and eager to accept it. Evolution was the only naturalistic explanation for the beginnings of life and remains so to this day. It did have its detractors, though, and many scientists such as Agassiz, Pasteur, Mendel, Lister, and others actively opposed it. Still, evolution prevailed, not because it was scientifically sound but because it seemed to have a certain intuitive feeling that it was right. After all, doesn't similarity in form imply common ancestry and doesn't the variation in species prove that species are not immutable but change over time? Unfortunately, how magnanimous or exquisite a theory may seem to be does not make it a fact. Intuition and elegance of a theory are not scientific adjectives; science requires much more. Yet despite the lack of scientific proof, evolution was and has been accepted.

As mentioned, Darwin was never dogmatic about his position, and understood its theoretical basis. Darwin himself was actually remembered as a kind and gentle man. He was greatly bothered by the fact that his wife, a pious Christian, could never accept his theory. In a letter to his wife, he wrote, "When I am dead, know that many times, I have kissed and cried over this." Darwin was a man of great integrity, especially in scientific matters and was acutely aware that the whole thesis he had constructed in the *Origin* was entirely hypothetical. Further, Darwin knew that the lack in the fossil record of intermediates, proposed to him by Agassiz and others, posed a great problem. "The distinctiveness of specific forms and their not being blended together by innumerable transitional links is a very obvious difficulty."

Darwin rejected the idea of saltationalism, which are sudden leaps in evolution. Darwin's doubts concerning the lack of fossil evidence eventually became Darwin's hope that transitional forms in the fossil record would be found.

The theory of evolution had a profound effect on the world. The evolutionary thesis implied an end to traditional theologic and anthropologic thought. Man was no longer the pinnacle of creation but instead a "cosmic accident" that resulted from purely random processes, no different from any of the other creatures on Earth. Thomas Huxley, a wealthy devout agnostic and strong opponent to all organized religion and a contemporary of Darwin who described himself as "Darwin's Bulldog," understood the implication of Darwin's theory and saw no compromise between science and religion. Richard Dawkins, Darwin's current bulldog, in his book, *The Blind Watchmaker*, writes, "Darwin made it possible to be an intellectually fulfilled atheist." Dawkins goes on to say, "It is absolutely safe to say if you meet somebody who claims not to believe in evolution, that person, is ignorant, stupid, or insane (or wicked, but I'd rather not consider that)." Ironically, Dawkins says what he dislikes most about creationists is their "intolerance"!

Michael Denton understands well the implications of evolution and its effects on society. Again, from his book, *Evolution: A Theory in Crisis*, he writes, "Chance and design are antithetical concepts and the decline in religious belief can probably be attributed more to the propagation and advocacy by the intellectual and scientific community of the Darwinian version of evolution than to any other single factor." He further writes, "It was because Darwinian theory broke man's link with God and set him adrift in a cosmos without purpose or end that its impact was so fundamental. No other intellectual revolution in modern times (with the possible exception of Copernican) so profoundly affected the way men viewed themselves in their place in the universe."

It has now been over 150 years since the publication of Darwin's book, and evolution remains as much a theory now as it was then. The "innumerable transitional links" that Darwin knew must be found to prove his theory have not been found, and despite the assiduous efforts and despite the vitriol of eminent atheists such as Dawkins and others, there is simply no proof of molecules-to-man evolution, and just saying it over and over and over again does not make it so.

Chapter 3

Defining Terms

Evolution, Kinds, Taxonomy, and Geologic Column

One of the problems in discussing evolution involves the confusion of terms. When a scientist uses the word "evolution," his or her meaning of the word may or may not be what you and I are thinking because evolution is a very broad term. Because of this confusion, as well as confusion over other terms and concepts, it is necessary to define them as precisely as we can.

Evolution

One of the best and most thorough definitions of evolution comes from Ker C. Thomson, formerly Professor of Geophysics at Baylor University, in the book *In Six Days*. "Following the laws of physics and chemistry, the concept is that through 'natural selection' operating over vast periods of time, fortuitous favorable events happened that brought about successively more complex biological chemicals which again, either fortuitously or through some undefined property of matter, concatenated (the ability of an element to form bonds with other elements, uniting atoms to atoms), leading upward to protocells, cells, living creatures, and then man himself." A more concise yet appropriate definition of evolution comes from Bolten Davidheiser. He states, "Evolution means that all of life on earth developed from one or few simple life forms. These alleged simple ancestors developed in a natural way from non-living matter."

So for our purpose we will be using these definitions of evolution that I will summarize as "molecules-to-man evolution," which conforms to the "general theory of evolution" proposed by Dr. Michael Denton. This is extremely important and will be a recurring theme. Do not be fooled by evolutionists who want to make synonymous the general theory, or macroevolution, with the special theory, or microevolution (adaptation).

Microevolution describes small changes in species and no one really doubts that occurs. If that were the only definition of evolution, then call me an evolutionist because change over time within species is observable and undeniable. Genetic variability is obvious within kinds. Compare, for example, canines. There are massive dogs like the Great Dane and very small ones like the Chihuahuas. Canines include the wolves, the foxes, and coyotes as well. Variability within *Homo sapiens* or humans is also clearly obvious. Once again, human beings come in all sizes and shapes, different colors, and varying degrees of other characteristics. It is self-evident that species change or "evolve" if you must, but there is a great difference between the special theory and the general theory, which allows for life to begin from non-life and morphologic kinds to change into other kinds, for which there is little evidence. In fact, when evolutionists allude to thousands of examples of transitional forms in the fossil records they are really referring to changes within species. This, however, brings up yet another question, and that is how to define species.

Kinds
Today, there are at least twelve different definitions of species so whoever you are talking with must understand how you are defining the term. The word "species" itself came from the Latin rendering of the Hebrew word for "kind" found in the book of Genesis. This is somewhat unfortunate, as it has caused some confusion throughout the ages. At one time, the Genesis concept of kind was our concept of species. The word "species" now conveys a different meaning.

The words *bārā' mîn* is a term now used for the study of kinds, also termed *baraminology*. This is derived from the Hebrew word *bārā'* meaning "to create" and *mîn* meaning "kind." The Hebrew word *mîn* or "kind" is found in three places in the book of Genesis. In Genesis 1, God creates plants and animals "according to their kind," and in Genesis 6 and 8 God instructs Noah to take "kinds" of all animals into the Ark and for these "kinds" to reproduce after the Flood. As can be seen, animals were to reproduce within the confines of their kind. Though hard to necessarily define, kind certainly may be said to include animals that were able to breed together. For example, we don't see dogs mating with cats or horses mating with cows. Keep in mind, two animals can produce a hybrid (a horse breeding with a donkey produces a mule) and still be considered kinds. It is clear as the terms are used today, species and kind are not synonymous. Today, "species" is a definition by man

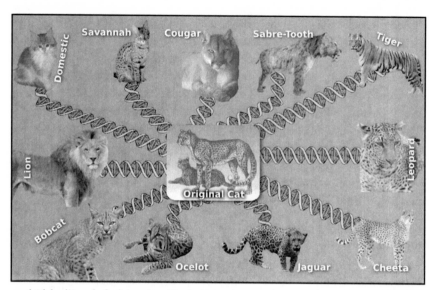

and "kind" a definition by God. Certain species can breed together and produce offspring. For example, a male lion can breed with a female tiger to produce a liger, the domestic cat has been bred with a lynx to produce a lynxcat, and zebras have mated with donkeys to produce zonkeys (see photo above). These, however, do not produce different kinds. Most of these creatures cannot reproduce themselves (although some can). So from my viewpoint, just the fact that organisms cannot reproduce does not necessarily infer a new kind and this is important. Kinds may more properly be defined as "family" in man's taxonomy but we have to be very careful using man's divisions. The biblical term "kind" cannot be defined precisely by manmade terms. In fact, we may never be able to fully delineate all the kinds.

As mentioned before, the terms "micro-evolution" and "macro-evolution" are often used to describe small changes in organisms vs. large ones. Natural selection clearly allows for adaptation of organisms, the most classic example coming from the finches at the Galapagos Islands. These finches adapted on the various Galapagos Islands based upon their sources of food and other factors. We will look at this in more detail later but it is very important to understand the difference between adaptation and evolution. Animals and plants adapt, but there is no evidence of change in kinds.

Taxonomy
Taxonomy is the discipline in science that classifies organisms into groups

based upon similarities. The Linnaeus Classification System, named after Carolus Linnaeus, a Swedish botanist, who lived in the 17th century, is still considered by most the best classification system today and is widely used. His hierarchical system has seven levels and they are, from the smallest to the largest; Species, Genus, Family, Order, Class, Phylum and Kingdom. For an example, let's examine the grizzly bear, also known as *Ursus arctos*. This name is its genus and species name. It is of the family *Ursidae*, order *Carnivore*, class *Mammalia*, phylum *Chordata*, and kingdom *Animalia*. What must be emphasized is that all taxonomic classification systems are defined by man, not God. Man separates organisms as far as order or species. The Bible's "taxonomy" is *kind*.

The second important point to make is that the 18th and 19th century taxonomists, such as Carl Linnaeus, followed a typological method. Typology implied there were absolute discontinuities between each class of organisms and typologists acknowledged the existence of biologic variation but denied it could ever be radical or directional. For the typologists, each individual member of a class conformed in all essential details to an archetype (a theoretical and purely hypothetical entity).

If these groups or classes could be arranged lineally or sequentially, and connected together through a series of transitional forms leading back to one original source, it certainly would give much more credence for evolutionary theory. This, however, is clearly not the case. Intermediates are virtually unknown, and as Michael Denton has stated, "It is impossible to allude to any more than a handful of cases where the pattern of nature seems to exhibit something of a sequential argument."

At the species level, typological differences can be hard to delineate but at levels above the species, typological models hold almost universally and there is no reason to believe organisms were ever converted to another group. We shall see later how this is verified from the fossil record.

One final thought about taxonomy: many leading biologists of the 18th century rejected evolution. Men such as French biologist Georges Cuvier; American zoologist Louis Agassiz, Richard Owen, a British anatomist who coined the term "Dinosaur"; anatomist Asa Gray and many more simply found no compelling evidence for a sequential pattern in nature. They did so not based on a theological argument but on the lack of empirical evidence for the theory. As Darwin responded to Asa Gray, "One's imagination must fill up the very wide blanks."

Geologic Column

According to the high school textbook, *Biology,* by Kenneth R. Miller and Joseph S. Levine, "Paleontologists use divisions of the geologic time scale to represent evolutionary time. The geologic time scale is based upon the concept of the geologic column." Evolutionists describe the geologic column as a vertical line of many different rock formations that represent every year of the earth's history to this date. The first rocks in the column were laid down about 4.5 billion years ago; however, fossil remains only began at least one billion years later, with multicellular life forms 650 million years ago, per the geologic clock. Supporters of the column and evolution say organisms slowly evolved during each period, finally culminating in what we have today. Each layer of the column above the Hadean Eon has fossil remains of animals and plants, most of which are extinct, and these fossil remains help to identify the column. These fossil remains are given the term "index fossils." Fossils are then identified by the rocks they are in and then are used to date the rocks!

Ironically, the geologic column was devised before 1860 by catastrophists and creationists, Adam Sedgwick, Roger Murchison, William Conebere, and others, to affirm the earth was primarily formed by catastrophic events. This clearly is not how the geologic column is accepted today.

The geologic column is used to describe eons, eras, periods, epochs, and ages in time. The Hadean Eon represents time before life and before fossils are found, and represents approximately 4 billion years of time, the earth's age given between 4.5 and 5 billion years. Fossils are found in the Phanerozoic Eon. The oldest era in the Phanerozoic Eon is Precambrian time with only one period, the Vendian, the last era being the Cenozoic with the Quaternary period (present time). The Precambrian Era has very little fossil remains and yet occupies 88% of the earth's history after the Hadean Eon, according to the geologic timescale (see geologic column, next page, although the eons are not included in that chart).

But is the geologic column real or just a hypothetical construct? First consider a portion of the geologic column, the Phanerozoic Eon, is supposed to be 100 to 200 miles thick and should be found everywhere since it represents the entire timeline for that eon, yet it does not appear all over the earth. In fact, it is apparently found in only 50 locations which represents about 1% of the earth, and if you add the ocean floors it only represents 0.4% of all the places the column should be. As an example, the Grand Canyon, one of the oldest, deepest land exposures in the world, contains less than one-half of the column.

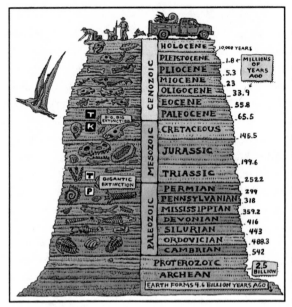

Apparently nowhere does the entire column exist in its full thickness of at least 100 miles. The average thickness is about one mile and in a few places it is sixteen or so miles in depth. Clearly, the geologic column as depicted in most of the textbooks exists nowhere. Also, there are hundreds of locations where the order of the strata does not match the order of the geologic column.

These are just a few problems with the geologic column, but suffice it to say the concept of geologic time and geologic column is largely a hypothetical construct. The relatively few places where the twelve periods exist in the Phanerozoic eon (the eon when life is supposed to have begun) can be explained by both old earth and young earth adherents. Later in our study we will look at young earth vs. old earth evidences, but for our current purposes it doesn't really matter. This is because no matter how many millions of years you assign to the age of the earth, the evidence for evolution is still meagre. Evolutionists believe that given enough time the impossible becomes possible. I believe that time does not help them. Evolution is not true regardless of the age of the earth. We will from time to time refer to the geologic column regardless of whether it has merit or not, because the column that evolutionist think proves their theory actually refutes it!

Chapter 4

Homology

The theory of evolution fails or survives on the validity of the notion of common ancestry proposed by Darwin in his book, *On the Origin of Species*. Obviously, human eyes have never witnessed this, and Darwin knew the fossil evidence did not support his common ancestry idea; therefore, for Darwin the most compelling evidence came from the similarities of structure many animals share. This similarity is known as *homology*, first proposed by Richard Owen in 1840. For Owen, similarities of structure, or homology, and similarity of function, or analogy, were evidence of an archetype, or a common designer or plan. Other biologists such as Louis Agassiz, shared this concept with Owen. Darwin, though, adulterated this to describe homology as evidence for common ancestry.

It cannot be overstated how important homology was to Darwin, and not only Darwin but for evolutionary biologists ever since. This concept is alluded to over and over again in *Origin*. The fact that vertebrate animals, especially the mammals, shared similar structure and body types, were symmetrical with two eyes, two forelimbs, and two hind limbs, shared similar organs and systems such as the neurological system with brain and spinal cord, digestive system, cardiovascular system, respiratory system, and reproduction system were to him strong evidence for evolution. A few excerpts from Darwin's *Origin* will highlight the importance he saw in homology. "We have seen that the members of the same class, independently of their habits of life, resemble each other in the general plan of their organization ... is it not powerfully suggestive of true relationship of inheritance from a common ancestor?" Also from *Origin*: "What can be more curious than that the hand of a man, formed for grasping, that of a mole for digging, the leg of a horse, the paddle of a porpoise, and the wing of a bat, should all be constructed on the same pattern, and should include the same bones, in the same relative positions?" Darwin was alluding to the so-called pentadactyl pattern seen in all the major vertebrates. This pentadactyl pattern, at least on the surface, could be explained by a common ancestor. Would a creator be limited to this design? Darwin considered it a hopeless cause for creationists. "Nothing can be more hopeless than to attempt to explain this similarity of pattern in members of the same class, by utility or by doctrine of final cause." That animal's share similar structures is evident. For creationists, homology is not the question, the question is the cause!

The phenomenon of homology has been used ever since as a mainstay in evolutionary doctrine and is found in all biology textbooks and in encyclopedias. But is common ancestry the only or even the best explanation for homology? First though, let's look at how "homology" is now defined. *Webster's Dictionary* defines it as "a similarity often attributable to common origin; likeness in structure between parts of different organs due to the evolutionary differentiation (as the wing of a bat and the human arm) from a corresponding part in a common ancestor." Biologists define it as, "the existence of shared ancestry between a pair of structures or genes in different species; descent with modification from a common ancestor." Do you notice anything peculiar about these definitions of homology? Obviously the definitions are circular, that is, they beg the questions with the conclusions buried in the premises. They clearly presuppose common ancestry. In other words, they use the definition for evidence when the definition already assumes the evidence is true. To this point, Darwin's definition of homology is when, "A structure is similar among related organisms because those organisms have all descended from a common ancestor that had the equivalent trait."

So Darwin and subsequent evolutionists have posed the question, "If every organism were created independently, it is unclear why there would be so many homologies among certain organisms while so few among others?" For men like Richard Dawkins, evolution "works" by modifying pre-existing structures. To the evolutionists, this makes perfect sense, but of course it does not explain how the pre-existing structures got there to begin with.

All evidence for homology is then deductive and based upon some supposed scientific "elegance." In *The Descent of Man*, Darwin based his whole theory of man's descent from apes on this reasoning, relying solely on the homology argument. For Darwin, common descent was the only logical explanation.

Certainly the validity of an evolutionary explanation for homology would be strengthened if embryonic development were specified by homologous genes. Also, one would expect that if a group of animals came from a common ancestor then the mode of germ cell formation would also be essentially the same from one animal species to the next. In other words, if homologous structures developed from similar genes on chromosomes of organisms with similar germ cells this would at least help explain from an evolutionary point of view how these structures developed from one organism to the other. This would at least suggest a "true relationship" but as Michael Denton has pointed out, "It is becoming increasingly clear that the principle cannot be extended in this way." This is because in nature homologous structures are frequently specified by non-homologous genes and the concept of homology "seldom extends back to embryonic development." Also, sometimes we see the opposite where non-homologous genes produce similar structures. Furthermore, the mode of germ cell formation varies significantly from one animal species to the next.

Biologists in fact have known for some time that homologous structures are not due to similar genes yet the argument continues because it is so compelling to evolutionists. Evolutionists employ somewhat circular reasoning in defining homologies as structures that are derived from a common ancestor as we have already noted. Ernst Mayr, one of the principal architects of Neo-Darwinism, states, "After 1859 there has been only one definition of homologous that makes biological sense ... attributes of two organisms are homologous when they are derived from an equivalent characteristic of a common ancestor." Once again, clearly, an example of circular reasoning; you cannot use homology for evidence for evolution and use it as the explanation for it as well.

This homology process was especially evident to Darwin when he looked at the four limbs of vertebrates, specifically the wing of the bat, the flipper of the porpoise, the

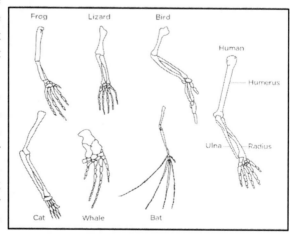

front leg of the horse, and the arm of a man. For Darwin, this was proof of homology and also proof of common ancestry. This, as pointed out by Jonathan Wells, is "putting the cart before the horse" and as Alan Boyden in 1947 wrote, Homology requires "that we first know the ancestry and then decide the corresponding organs and parts are homologous." Biologists themselves know that similarity does not always make evolutionary sense. For example, the eye of the octopus and the eye of the human structurally are very similar but no one, even the evolutionists, argues that humans and octopi have a common ancestor!

In *Icons of Evolution*, Jonathan Wells quotes philosopher Ronald Brady, in 1855, "By making our explanation into the definition of a condition to be explained, we express not scientific hypothesis but belief. We are so convinced that our explanation is true that we no longer see any need to distinguish it from the situation we are trying to explain."

Some have tried to use the fossil record to give evidence of common descent by identifying homologous structures, but this has failed as well because, once again, it falls back on circular reasoning. It is one thing to call something "eloquent" and "exquisite," but it's quite another to have scientific proof for one's theory. If there is proof for homology it will be found in the fossil record, since clearly we cannot observe it. But in reality the fossil records offer no such proof and where examples are given by evolutionists, they are always examples of micro-evolution. For example, the trilobites of the Ordovician period found in the fossil record dated 500 million years ago are given as proof. The trilobites "evolved" during this period, with the number of ribs changing, and as a result some biologists even classify these "new trilobites" into a different genus. But no matter into what genus man places these, they are still trilobites; and of course no one proposes where the trilobites themselves evolved from to begin with, a topic we will discuss later. Also Niles Eldredge's Ph.D. work was on the evolution of the trilobite eye which never changed over time and was always fully developed and more advanced than the human eye in some ways. Other examples could be cited but all of them are inferences and offer no plausible "proof."

In 1990, Tim Berra defended Darwinian evolution by comparing the fossil record to a series of automobiles, specifically the Chevrolet Corvette. Berra looked at how the Corvette had changed over the years and stated this was similar to how evolution had occurred in animals. By drawing this analogy, however, Berra was really showing that the change seen in organisms

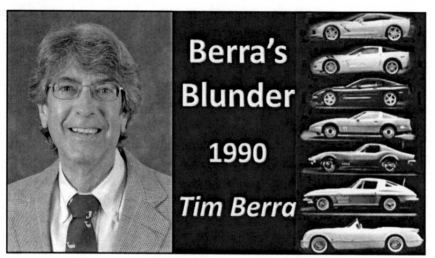

Berra's
Blunder
1990
Tim Berra

through the fossil record, or for that matter that of a Chevrolet Corvette, could more easily be explained from a creator or intelligent design point of view. We know that the Corvette did not "evolve" and this argument has been termed "Berra's blunder" by law professor Phillip Johnson, in his book, *Darwin on Trial*. Showing that organisms share similar structures gives absolutely no proof for evolution; for that it is necessary to show a biologic process that could have caused this, otherwise "descent" and "modification" are just words.

As stated before, evolutionists have tried to show that homologous structures are produced by similar developmental pathways in the embryonic state but this does not fit the evidence. First, from the very beginning of cell division, it is clear from the blastula stage that organisms, even within the same order, are quite different. Without going into great scientific detail, it is clear there is great dissimilarity in early embryogenesis in the different vertebrate classes and homologous structures cannot be traced back to homologous cells. For British embryologist Sir Gavin de Beer, apparent homologous structures formed during embryogenesis may not be homologous at all! There is another problem embryologically as well when it comes to homology, and that is that almost every gene structure in higher organisms has been found to affect more than one organ system; this effect is known as pleiotropy. It is also clear that the pleiotropic effect is species specific.

Similarity in structure and function then are obvious to both the creationist and the evolutionist, but the question then becomes: What does this prove? The requirements for life are similar for living organisms and preferred designs are so because they do the function they are designed to do and work! Tires are round because they work well on a car, bicycle, wheelbarrow, motorcycle, etc. Kidneys work because they are suited for their function whether it be in the body of a bear, a horse, or a human.

Yet another problem with homology is the similarity of function but not structure that is seen in the animal world, so-called analogy. Take flight, for example. Birds fly, insects fly, mammals fly, and even fish fly to some extent, but their structures for such flight are quite different, meaning flight would have to have evolved in different animals independently, i.e. convergent evolution forming "analagous organs." Evolutionists call this convergent evolution but this is no explanation at all, rather just a naming of the process. Even the example of the pentadactyl design has problems. As pointed out, the forelimbs of terrestrial mammals follow this design but so do the hind limbs, yet as Michael Denton points out, "No evolutionist claims that the hind limb evolved from the forelimb, or that the hind limbs and fore limbs evolved from a common source." How this complex and seemingly arbitrary pattern could evolve twice independently is "mystifying." Unity of type is just not readily explained by evolutionary theory.

Variation within species is problematic for the Darwinist, as are the limits imposed on organismal variation. For example, what advantage does variety give to an organism? Take humankind for example. What advantage is it for one person to be able to roll the tongue and another not; or for one to have an attached earlobe and the other one not; or a hitchhiker's thumb, widow's peak, or bent little finger? There is simply no evolutionary advantage offered by these traits humans possess. Evolutionists may try to explain these traits as vestiges of structures that once had more of an evolutionary advantage, but again this is no more than a hypothesis for which there is no proof. As we have seen, organisms all show tremendous variation within each kind. It is clear, however, that when excess in deviation occurs the animal's ability to survive is greatly compromised. Large mutations occur only at the demise of the organism.

So what does unity of type or homology point to, if it's not evolution? How about a common designer! If God chose DNA to be the underlying blueprint for all life, why should it be surprising that He would use similar homologous structures in His design of animals? In fact, it would be much harder to believe that a common designer would refrain from using similar features in the *ex nihilo* creation of thousand of varieties of animals. What separates man from animals is not flesh or DNA but spirit. Only in man was the spirit or soul given, and it is in this manner by which we are made in his likeness. Genesis 1:27 says, "And God created man in his own image, in the image of God he created him; male and female he created them."

For Darwin, homology was virtual proof for evolution but that same homologous resemblance really helps distinguish one class from another unambiguously. At best, homology gives only circumstantial evidence for evolution and circumstantial evidence can be quite deceiving. I found quite enlightening the quote from Conan Doyle's Sherlock Holmes in Denton's *Evolution: A Theory in Crisis.* "Circumstantial evidence is a tricky thing, answered Holmes thoughtfully, it may seem to point very straight to one thing, but if you shift your point of view a little you may find it pointing in an equally uncompromising manner to something entirely different. There is nothing more deceptive than an obvious fact."

How about Haeckel's Embryos?

In *On the Origin of Species* Darwin wrote, "It seems to me the leading facts in embryology, which are second to none in importance, are explained on the principle of variation and the many descendants from some one ancient progenitor." Continuity of embryonic structure reveals continuity of descent. The phrase "Ontogeny Recapitulates Phylogeny" was later coined for this concept by Darwin's German Bulldog biologist, Ernst Haeckel. Haeckel referred to this as his "biogenic law." From this biogenic law, Haeckel proposed that "on its way to birth a human becomes a fish, an amphibian, and so on up the evolutionary ladder." Haeckel used the term "ontology" to designate embryonic development and "phylogeny" to designate the evolutionary history of the species. In other words, during the development of the organism the embryo goes through the various evolutionary stages that predated it.

For evidence of this, Haeckel looked at the embryos of fish, salamanders, tortoises, chicks, hogs, calves, rabbits, and humans. Their appearance at what Haeckel called "the earliest stages" look quite similar and only diverged later in development. But is this true and if so what does it prove? First, it should be noted that Haeckel purposely chose embryos that did look similar and his drawings depict only five of the seven vertebrate classes for this reason. Secondly, and not insignificantly, Haeckel faked his drawings! In other words, Haeckel changed the appearance of his drawings and later his wood carvings to look more similar than they really were. Thirdly, what Haeckel labeled the "first" stage of embryonic development in his drawings was no such thing; it was actually midway through it. We have already noted

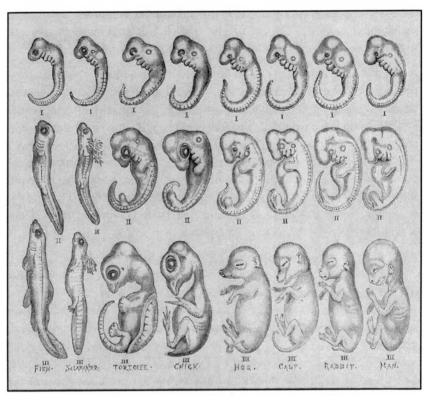

FISH. SALAMANDER. TORTOISE. CHICK. HOG. CALF. RABBIT. MAN.

that at the blastula stage, organisms are quite different in appearance, even to a not so trained eye. So if at their earliest stages, embryos are very dissimilar, this would cause one to reason more appropriately that they do not have a common ancestor but are of separate origins (see chart next page).

Now there is a time that embryos do have some similarity in appearance, but this is midway through the development, not the "first stages" From this midpoint stage, organisms diverge and become increasingly different until their "adult stage." Rudolph Raff calls this the "developmental hourglass." What is clear is that there is no evidence of organisms recapitulating their evolutionary history. Embryos do not go through evolutionary stages and humans do not have gills! The gills Haeckel painted into the embryos are "brachial arches" that go on to form structures of the neck area such as the pharynx, and never function in any respiratory capacity at all. In fact, these same "brachial arches" or "clefts" or "pouches" once purported to be gill slits don't ever become gills, even in fish!

What remains surprising, or maybe not so surprising, is that textbooks continue to refer to Haeckel's embryos and continue to refer to ontology recapitulating phylogeny, although frequently attaching Karl Ernst von Baer's name to it, apparently in an attempt to avoid the embarrassment associated with Haeckel vis-à-vis his forgeries.

To prove the point, this comes straight from a biology textbook edited by Peter Raven and George Johnson in 1999. "Notice that the earliest embryonic stages of the vertebrates bear a striking resemblance to each other. Some of the strongest anatomical evidence supporting evolution comes from comparison of how organisms develop. In many cases, the evolutionary history of the organism can be seen to unfold during its development with the embryo exhibiting characteristics of its ancestors."

Some, such as Douglas Futuyma, a professor of evolutionary biology and author of a graduate textbook continue to purport the presence of gill slits in vertebrate embryos and the similarity of appearance of bird and mammalian embryos. Apparently Professor Futuyma did not know of Haeckel's forgery until it was pointed out to him by a creationist! The point is that homology, embryology, and Haeckel's embryos continue to be used in college and high school textbooks as evidence for evolution; but as we have seen, only through circular reasoning can that fly.

Chapter 5

DNA and the Genome

During Darwin's life very little was known about genetics, and certainly virtually nothing about DNA and the information-carrying capacity it contains. As a result, as we have seen, homology was the prime evidence Darwin gave for evolution, realizing the fossil evidence was slim at best. Almost 100 years after the publication of *On the Origin of Species*, the chemical structure of DNA was discovered; and very shortly thereafter Francis Crick's sequence hypothesis was realized, which confirmed the immense information storage capacity of this intriguing molecule. Some fifty years later, the entire human genome was "decoded."

These significant discoveries now offered supposed proof to the Darwinists, independent of homology and the fossil records, that evolution was a necessity. For Francis Collins, a medical doctor and geneticist, and head of the Human Genome Project, looking at human DNA and what lies between the genes, plus the findings of similar sequences on the genomes of other distantly related organisms, gave "powerful support" to Darwinian Theory. For example, the order of the genes along the human and the mouse chromosomes is generally maintained over substantial stretches of DNA. This gives evolutionists "proof" of common ancestry. Further evidence for the evolutionists comes from the imperfection of the genome, which is "riddled with useless information," referred to as "junk DNA." As a result, Philip Kitcher called the Intelligent Design movement a "dead science" unable to explain the new evidence from genomic studies. One scientist even referred to the intelligent design scientists as "Creationists dressed up in tuxedos."

But is it true that genetics and the genome have given a death blow to creationism? Further, does the intelligent design community have no answers concerning the evidences of evolution in the genome? We will answer these questions subsequently, but first it is necessary to examine DNA and how important it is to understand its molecular design in answering the evolutionists' arguments. Before we do that, I think it is interesting to look at the discovery of DNA and how the structure was finally identified, discoveries which indeed have had profound implications. Christians need not fear the science of DNA and to do so is simply turning a blind eye. Albert Einstein once said, "Science without religion is lame, religion without science is blind." Galileo, centuries before, said, "I do not feel obliged to believe that the same God who endowed us with sense, reason, and intellect has intended us to forego their use."

The Discovery of DNA

James Watson and Francis Crick

Most people have the belief that James Watson and Francis Crick discovered DNA in 1953, but this is technically not true. Scientists have known about DNA and its chemical makeup since the discovery of "nuclein," later to be called DNA, by Swiss biologist, Friedrich Miescher in 1869, ten years after the publication of Darwin's book. Obtaining bandages from a local hospital, Miescher examined white blood cells, eventually extracting the gooey substance from the nucleus, which he referred to as "nuclein." The exact chemical makeup was not finally known until 1909. However, the structure of the DNA molecule remained a mystery, becoming one of the greatest pursuits in scientific history. What became apparent was that the molecule of DNA, no matter what organism it came from, always had equal amounts of the nitrogen bases, adenine, and thymine as well as guanine and cytosine, a finding attributed to Erwin Chargaff at Columbia University.

By the mid 20th century, shortly after World War II, there were essentially three teams of scientists feverishly working on the structure of DNA. There was Linus Pauling and his group in the United States, Lawrence Bragg and Max Perutz at Cambridge, and Maurice Wilkins and Rosalind Franklin at King's College in London, all trying to be the first to put the structure of DNA together. Contrary to what you might think, scientists are pretty selfish and rarely share their work with one another, and this was certainly the case in the discovery of DNA. Had they shared their knowledge, undoubtedly the discovery of its structure would have occurred much sooner than it did.

Also working at Cambridge during this time were a young biologist from Chicago University, only recently having received his Ph.D., 23 year-old James Watson, and an equally young physicist, as yet to earn his Ph.D.,

Francis Crick. Fortuitously for them they became aware of Rosalind Franklin's work with x-ray diffraction. Using x-ray diffraction, Dr. Franklin was able to visualize the DNA molecule and realized it "may be a double helix" but it certainly was not a triple helix as had been previously postulated. Seizing on this x-ray data, Watson and Crick, two most unlikely "heroes" began putting the structure together, already knowing DNA's chemical makeup. It became apparent to them that the DNA molecule was arranged with a five-carbon-based sugar, deoxyribose, and the phosphate group on the outside, and the nitrogen base, held to the sugar group by a hydrogen bond on the inside. These nitrogen bases were then connected to one another in a "strand-like fashion" of the DNA molecule, with adenine always attached to thymine and cytosine always attached to guanine by hydrogen bonds, in a "double helix manner." They quickly published their seminal work in the *Journal of Nature* in 1953 with the title, "Molecular Structure of Nucleic Acid: A Structure of Deoxyribonucleic Acid," giving little credit to the work done by Rosalind Franklin. Watson and Crick would eventually receive the Nobel Prize in 1962 but Rosalind Franklin would not, having died at age 37 of ovarian cancer, possibly as a result of her exposure to radiation (Nobel Prizes are never given posthumously).

At the time of their discovery, Watson and Crick did not realize its full implications. Later, Crick offered the "sequence hypothesis" on how DNA coded information. His hypothesis was later confirmed and as a result the structure of DNA would forever change our understanding of life. Life was not just about matter and energy but information! We now understand that DNA is the source of "genetic information," the "genetic message," the "genetic blueprint," "assembly instructions," and the "digital code."

The Molecule

Webster defines information as "the communication or reception of knowledge or intelligence." To understand how DNA functions, we must look at its structure and see how information is transferred. Knowing this will help us realize how unlikely it is that DNA would have just appeared spontaneously and "naturally." As Dr. Collins has put it, "No current hypothesis comes close to explaining how in the space of a mere 150 million years the prebiotic environment that existed on planet Earth gave rise to life." Further, Stephen Meyer says, "Evolutionary theory could not explain the origin of the first life because it could not explain the origin of the genetic information in DNA." Finally, from James Hodges' book, *Creation vs. Evolution,* comes this from John Sanford speaking of DNA, "Its very existence is a mystery. Information and complexity which surpasses human understanding are programmed into a space smaller than an invisible speck of dust. Mutation/selection could not even begin to explain this. It should be clear that our genome could not have risen spontaneously. The only reasonable alternative to spontaneous genome is a genome which occurred by design."

Before looking closely at the DNA molecule, first we need to understand that in and of itself, DNA is nothing special. DNA provides the substance for the information but it is not the information itself. George Williams, himself an evolutionary biologist, has said, "Evolutionary biologists have failed to realize that they work with two more or less incommensurable domains: that of information and that of matter. ... The gene is a package of information, not an object. The pattern of base pairs in DNA molecules specifies the gene but the DNA molecule is the medium, it's not the message." DNA has been referred to by Dr. Francis Collins as the "language in which God revealed life." Even atheist Richard Dawkins has alluded to how "the machine code of the genes is uncannily computer like."

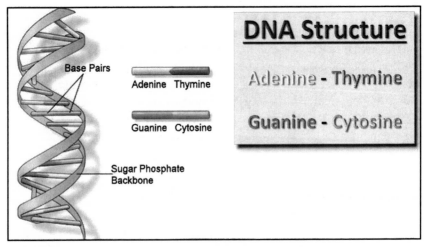

DNA Structure

Let's now take a look at this remarkable molecule, DNA. As we have noted, DNA stands for deoxyribonucleic acid. DNA is a polymer made up of nucleotides, hence a polynucleotide. It has three basic molecules, a five-carbon sugar (deoxyribose), a phosphate group, and nitrogen bases, guanine, cytosine, thymine and adenine as we have already seen arranged in a double helix structure. DNA is contained in chromosomes of eukaryotic cells that are further contained in the nucleus of these cells. It is a very long molecule, being 19 million times longer than its width. If you stretched the DNA in one cell of a human being, the DNA would stretch over 6 feet tall. Since our bodies have an estimated 50 trillion cells, if you could line up the entire DNA from just one person it would stretch to the sun 600 times! DNA is also a very fragile and unstable molecule and would disintegrate quickly if left by itself. DNA contains a 6-billion letter code. The human chromosome number 1 has 270 million base pairs whose representative letters would fill 200,000 pages. These "codes" store information for what goes on in our bodies. A gene is a discrete segment of DNA that codes for a particular function, especially for the production of protein, by lining amino acids in a sequence, the average sequence being 100 to 150 amino acids. But how is this done?

This will of necessity be an abbreviated version; but first consider that DNA, through the help of RNA, has the unique ability to replicate itself. RNA is similar to DNA but is single stranded, contains ribose as its sugar instead of deoxyribose, and substitutes uracil for thymine in its nitrogen bases. Through an uncoupling effect, DNA separates from its double helix structure into a single rather than double strand, exposing the nitrogen bases. Messenger RNA or mRNA then attaches to the DNA molecule, with uracil always bonding with adenine and cytosine always with guanine. Three base groups or codons then code for a specific amino acid. As we have seen, these sequences of amino acids determine which proteins are made. After the proper sequencing occurs, the RNA releases from the DNA, exits the nucleus and then travels to organelles in the cytoplasm of the cell, such as ribosomes, to produce proteins which must be folded in an extremely intricate and precise process in order for the protein to function. Mistakes can sometimes be made in this process but DNA has the uncanny ability to "proofread" itself and correct them. Nonetheless, rarely, mistakes persist which we would refer to as mutations, the significance of which will be explained later. This process explained above is the very sequence hypothesis Francis Crick had postulated. It's amazing how accurate his hypothesis was. This comes from Crick in 1958: "If a sequence of bases constitute a gene, if genes direct protein synthesis, and if specific sequences of amino acids constitute proteins, then perhaps the sequencing of bases determines the specific sequencing of amino acids." DNA then specifies proteins, and proteins are specified by shape and arrangement.

The million dollar question becomes, how did these digitally encoded and specifically sequenced instructions in DNA arise? "Simple regular geometrics plus rules of chemical affinity did not and could not generate the specific complexity present in functional genes," so says Stephen Meyer. Curiously, the production of proteins requires DNA but the production of DNA requires protein! Jacques Monod stated in 1971, "The code is meaningless unless translated. The modern cell's translating machine consists of at least 50 macromolecular components, which in themselves are coded in DNA. The code cannot be translated otherwise than by product of translations." Richard Lewontin and David Goodsell have asked, "Which came first, proteins or protein synthesis?" The late British philosopher, Sir Karl Popper, has written, "What makes the origin of life and the genetic code a disturbing riddle is this: the code cannot be translated except by using certain products of its translation. This feature of life is exceedingly difficult to imagine."

Francis Collins understands the improbability of DNA. "DNA with its phosphate sugar backbone and intrinsically arranged organic bases, stacked neatly on top of one another, and paired together at each rung of the twisted double helix, seems an utterly improbable molecule to have just happened – especially since DNA seems to have no intrinsic means of copying itself."

Information
To understand just how improbable this is, we need to look at information and information theory. Information requires specificity, order, and complexity, all characteristics DNA possesses. To illustrate, look at a binary code and the genetic code of a protein.

Binary Code:

```
01010111011010000110010101101110001000000110100101101110001
00000011101000110100001100101001000000100001101101111011101
01011100100111001101100101001000000110111101100110001000000
11010000111010101101101011000010110111000100000011001010111
01100110010101101110011101000111001100100000011010010111010
```

This sequence is not just a random array of characters but the first words of the Declaration of Independence (When in the course of human events...) written in the *binary conversion* of the American Standard Code for Information Interchange (ASCII).

Genetic Code:

```
AGTCTGGGACGCGCCGCCGCCATGATCATCCCTGTACGCT-
GCTTCACTTGTGGCAAGATCGTCGGCAACAAGTGGGAGGCT-
TACCTGGGGCTGCTGCAGGCCGAGTACACCGAGGGGTGAG-
GCGCGGGCCGGGGCTAGGGGCTGAGTCCGCCGTGGGGCGC-
GGGCCGGGGCTGGGGCTGAAGTCCCGCCCTGGGGGTGCGCGC-
CGGGGCGGGAGGCGCAGCGTGCCTGAGCGCAGCGCCCCAT-
GAGCAGCTTCAGGCCCGGCTTCTCCAGCCCCGCTCTGTGATCT-
GCTTTCGGGAGAAACC
```

This string of characters is not just a random assortment of the four letters A, T, G and C, but a representation of part of the sequence of genetic assembly instructions for building a protein machine – an RNA polymerase – in a living cell. (These examples were taken from Stephen Meyer's *Signature in the Cell*.)

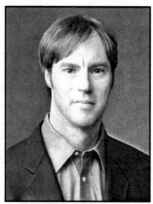

Stephen Meyer

Richard Dawkins has observed, there is indeed a similarity between a computer and DNA, both of which must have a programmer! Meyer has referred to this as the "DNA enigma," the mystery of the information needed to build the first organism. Charles Thaxton has suspected the process points to an intelligent cause, "the code in other words pointed to a programmer."

The complexity of DNA is not found anywhere else in nature. Some have pointed to crystals but crystals lack complexity and are not living. Others think polymers are similar to DNA but they lack specificity. Many have tried to determine mathematically how this complex molecule could have naturally arisen. By all accounts the likelihood is in essence zero and when we are talking about the origin of DNA we are in essence talking about origins of life.

Scientists must therefore answer three important questions when determining origins. First, where is the origin of the system for storing and coding information in the cell's DNA storage capacity; secondly, explain the origin of large amounts of specific complexity or functionally specified information in DNA; and thirdly, they must explain the origin of integrated complexity/the cell's information processing system. Since most scientists are not creationists this explanation must come from chance, law-like forces (necessity) or a combination of both. Many very smart and highly learned

individuals have looked into the likelihood of life originating spontaneously, and have concluded the likelihood is virtually impossible. Eugene
Wagner, Nobel Prize winner, believes the odds are overwhelmingly against
any process of undirected chemical evolution producing life. Professor Sir
Hermann Bondi, a mathematician who with Sir Fred Hoyle formulated
the concept of the Steady State Universe and who was also famous for his
work in relativity, rejected chemical evolution. Stephen Meyer points out,
creation of new information is typically connected with conscious activity.
Chance cannot "cause" anything and to use it as an explanation is to defer
to ignorance!

Ronald Fisher, a statistician in the 1920's, proposed a method of determining when chance can be explanatory. Taking examples from the gambling
world, Fisher plotted a "rejection zone" where chance becomes inconceivable. He stated, "The chance hypothesis can be eliminated precisely when
a series of events that deviates too greatly from the expected statistical
distribution of events based on what we know about processes that generate
those events or what we know from sampling about how frequently those
events typically occur." We know this intuitively. If a person in a Las Vegas
casino placed a bet on Red 16 100 times at a roulette wheel and won every
time, we understand how improbable this would be and the casino's bosses
would quickly investigate how the game was "rigged." Having said that,
the odds of rolling Red 16, 100 times is the same as the odds of rolling any
sequence 100 times, but what we recognize is a pattern. William Dembski,
a mathematician and author of *The Design Inference*, has pointed out, "The
odds of a sequence of 6,000 characters is roughly the same for any other
sequence of 6,000 but getting a pattern from that order is an entirely different
thing." For example, typing out in the computer randomly, "Four score and
seven years ago" and "neney taw jill.gnmek.hdx _," have the same likelihood and both contain information (what Dembski refers to as Shannon
information, named after MIT mathematician Claude Shannon); but clearly
one, the former, has meaning and the latter not. What is the difference?
Pattern! One suggests something other than chance and the other does not.

Is this important? Very much so! As an illustration, let's look at yet
another gambling example. Let's specify that to win the jackpot a gambler
must toss a coin that lands on heads 100 times in a row. As you can see, we
are specifying a precise pattern. So what is the likelihood? It is calculated
to be 1 in 1,267,650,600,000,000,000,000,000,000,000. Tossing the coin
continuously over one trillion years would not be enough time! Even still,

as Dembski says, "Scientists can never absolutely prove that some astronomically improbable event could not have taken place by chance. But they can determine when it is more likely than not that such an event wouldn't or didn't happen by chance alone." For this, Dembski has proposed an explanatory filter for the likelihood of an improbable event occurring. Using the total elementary particles in the universe at 10^{80} multiplied by possible interactions per second at 10^{45}, then multiplying this number by the number of events that could have taken place in the universe since its presumed origin at 10^{25} he derives a number of 10^{-150}, which he calls the "universal probability bound" of life occurring spontaneously. This is a number too large for us to really comprehend and it confers an impossible event.

Many others have tried mathematically to determine the likelihood of life occurring spontaneously. The most famous example comes from Sir Fred Hoyle who, in 1983, placed the odds of a single one-celled organism occurring by chance at $10^{40,000}$. The Wistar Institute, a "think tank" of scientists in 1966, looking at the origins of life, published an article, "Mathematical Challenges of Darwinism," in which they calculated the possibility of a short protein of 150 amino acids occurring by chance at 10^{195}. Robert Saver of MIT, an original part of the Wistar group, in the 1980's reported that the chance probability of achieving a functional sequence of amino acids was "exceedingly small," 1 in 10^{63} (keep in mind there are only 10^{69} atoms in our galaxy). Others have postulated the chance of just one functional protein coming forth spontaneously in some prebiotic soup as no better than one chance in 10^{164}. Finally, Douglas Axe of Cambridge, doing mutagenesis experiments, calculated the odds of a functional protein occurring by chance at 1 in 10^{77}. So what do all these astronomically high numbers mean? *They mean one has to throw away logic to believe life occurred spontaneously.* Using deductive reasoning and logic, Dembski has put forth his conclusion as a postulate. Premise 1: Life has occurred. Premise 2: Life is specified. Premise 3: If life is due to chance, life has a small probability. Premise 4: Specific events of small probability do not occur by chance. Premise 5: Life is not due to a regularity (regularities are natural laws like gravity). Premise 6: Life is either due to regularity, chance, or design. Conclusion: Life is due to design.

So why do many scientists still cling to the belief in molecules-to-man evolution? Some have tried to explain that life did not start with DNA but by RNA, but as Meyer has stated, "This does not solve the problem of biologic information but simply displaces it." Also, RNA is a very hard molecule to make but a very easy one to destroy, it possesses very few specific enzymatic proteins, and an RNA-based translation and coding system is implausible. The bottom line is, RNA doesn't solve the problem of naturalistic causes for the origin of life.

Some scientists, no doubt desperate to find naturalistic causes, have proposed DNA to be a result of "self-organization" and "biochemical predestination." This is simply wishful thinking not found in anything biochemically. DNA cannot be "explained" by just invoking bonding attributes, and Michael Polanyi has shown that even if living organisms function like machines they cannot be fully explained by reference to the laws of physics and chemistry. Origin of life experiments using chemical processes just do not produce large amounts of specified information!

One might appropriately ask at this point, how then can some scientists still hold to this molecules-to-man evolutionary theory? Quite frankly, most would say it simply does not matter! They will acknowledge the great improbability by invoking the "Anthropic Principle," which says that evolution has occurred as a fact regardless of its very high improbability. From their point of view, the fact speaks for itself.

At least Richard Dawkins has made an attempt to explain the origins of life naturalistically. As he states in his book, *The Blind Watchmaker*, he was able to program a computer to generate the sentence, "Me thinks it is like a weasel" spontaneously. But how did Dawkins do this? First, he programmed the computer to generate many separate "strings" (sequences of English letters). Then he programmed it to compare each string to the Shakespearean target phrase and selected only the string that most closely resembled the target. The computer then generated variant versions and compared them to the target over and over and over again until the target phrase, "Me thinks it is like a weasel" was produced. But do you see a problem here? This is not an example of a "blind" watchmaker. Dawkins' own intelligence was programmed into the computer. In fact, all computer-aided models that attempt to explain naturalistic causes for the origin of life suffer from the same generic problem; all are programmed by an intelligent programmer.

Some scientists accept the absurdity of molecules-to-man evolution but still cling to the overall notion of evolution. Francis Crick called it "a miracle." Klaus Dose, a German biochemist, in 1988 realized that the research efforts to date had "led to a better perception of the immensity of the problem of the origin of life on Earth rather than to its solution." At present, all discussions on principle theories and experiments in the field either end in a stalemate or a confession of ignorance. Even evolutionist, Stephen J. Gould, once concluded that humans are a "glorious accident which required 60 trillion contingent events."

For Francis Crick, the problem was solved by the idea of "panspermia." "Panspermia" is the concept that life originally was "seeded" from space. But this simply places the problem somewhere else, doesn't it? For Francis Collins, evolution occurred but the universe originated from God. Although a bit ambiguous in his book, *The Language of God*, Collins seems to believe that God either placed DNA in the world or created it here. In essence, Dr. Collins is evoking the theistic evolutionary concept. Why would he do this realizing how unlikely molecules-to-man evolution is? For Dr. Collins and other evolutionists, the answer lies in the human genome.

Chapter 6

The Genome

The genome of an organism is its genetic makeup or code. Chromosomes, meaning "dark bodies," contain DNA where genes are located in eukaryotic cells, and as we have seen genes are discrete segments of DNA, folded into the chromosome, that code for particular amino acids, which, after assembly and folding, are functional proteins. DNA represents 15% of the chromosome; the rest is made up of RNA and other proteins. Humans have 46 chromosomes (23 pairs), chimps have 48, dogs 78, butterflies 138, and one protozoan 1,600 chromosomes. Obviously, the number of chromosomes does not determine the complexity of the organism and neither does the number of base pairs. Humans have 2.9 billion base pairs, lungfish 130 billion, frogs 6.7 billion, and an amoeba 670 billion! The human genome, or our instruction book, was decoded in 2003 with other organisms being decoded soon thereafter. If one were to read one letter per second of our instruction book it would take 31 years to finish.

Francis Collins, Head of the Human Genome Project, points to several "surprises" found in our genome that were not expected. First, he points to how little of the genome is used to code for protein, it only being 1.5% of our DNA, which represents 20,000 to 25,000 genes (this number of genes was previously thought to be 100,000). The DNA not used for protein synthesis in our cells has come to be referred to as "junk DNA" for its apparent non-functionality. This junk DNA was thought by evolutionists to represent remnant DNA (sort of like vestigial organs). The second surprise for Dr. Collins was the number of genes in humans that were similar to those of other basic organ-

Francis Collins

isms such as worms, flies, and plants. Further, it was found that all humans are 99.9% identical. Only 0.1% allows for the great variation of men in all the races. The Genome Project finally related comparisons with our DNA to the DNA sequences of other organisms and found that when picking coding regions in human DNA (the part that codes for proteins) there will always be a significant match of the genome of other mammals. For Collins and others, this was powerful evidence for evolution. Also, for Dr. Collins, compelling evidence for a common ancestor came from the study of what is known as Ancient Repetitive Elements (AREs). These elements supposedly arise from "jumping genes" which were capable of inserting and copying themselves in various other locations in the genome without any functional consequence. He notes mammalian genomes are littered with such AREs and the human genome is made up of approximately 45% of these genes. He refers to these AREs as genetic "flotsam and jetsam." Much of the so called "junk DNA" contains these AREs. Dr. Collins and other evolutionists also find the similarities of the genetic structure of chimpanzees and humans to be evidence for common ancestry. Humans and chimps are identical by 96%, even though humans have 23 paired chromosomes and the chimps 24 (supposedly when humans evolved two chromosomes were fused together to generate one less pair than its ancestor, the chimp). Finally, evolutionists have looked at the genome and comparisons of DNA sequences from other organisms, and utilizing the concept of the molecular clock have constructed by a computer a tree of life which is similar to one Darwin proposed. For all these reasons Francis Collins and other evolutionists believe it is ignoring science to discount evolution at this point.

Still, Collins does not believe humans and chimps or apes are the same. Humans possess a "moral law" (a knowledge of right and wrong), language, altruism or "agape love," the ability to imagine the future, and a universal search for God. Apparently, Collins believes these attributes come from God as he has stated, "The comparison of chimp and human sequences, interesting as it is, does not tell us what it means to be human." When and how these traits were given is not revealed by Collins, who reports to be a Christian and certainly believes in the historicity of the New Testament. His explanation for the Genesis record apparently falls back on the gap theory and the use of figurative language.

But do Collins and other evolutionists have it right? Is the human genome "riddled" with useless information, as stated by biologist Ken Miller? Ken Miller argues, "The critics of evolution like to say that the complexity of the genome makes it clear that it was designed. ... But there is a problem with that analysis, and it's a serious one. The problem is the genome itself; it's not perfect. In fact, it's riddled with useless information, mistakes, and broken genes." So, can natural selection and mutations really explain how organisms have evolved from common ancestry?

To answer these questions, much of what I will refer to henceforth comes from Dr. Stephen Meyer's brilliant books, *Signature in the Cell* and *Darwin's Doubt.* First, let's look at the concept of "junk DNA." Again, junk DNA, or non-protein producing genes, are supposed to be left over from evolution. This concept cannot be overstated. When the human genome project released its findings, such men as Richard Dawkins, Philip Kitcher, Michael Shermer, and many others believed this dealt creationism a serious blow. As Philip Kitcher put it, "If you were designing the genomes of organisms, you would not fill them up with junk." In other words, why would God put DNA in chromosomes that had no function?

This is a valid point, as William Dembski explained and predicted in 1998, "On an evolutionary view you would expect a lot of useless DNA. If on the other hand, organisms are designed, we expect DNA, as much as possible, to exhibit function." Stephen Meyer points out, "The discovery in recent years that non-protein coding DNA performs a diversity of important biologic functions has confirmed this (Dembski's) prediction." In other words, there is no "junk DNA!" We now know non-protein producing DNA performs many functions including the regulation of DNA replication, regulating transcription, marking signs for programmed rearrangements of genetic material, influencing the proper folding and maintenance of chromosomes, controlling the interactions of chromosomes with nuclear membrane, controlling RNA processing, editing, and splicing, modulating transcription, regulating embryologic development, repairing DNA, and aiding in immune-defense or fighting disease. That's a lot of functions for "useless or junk DNA." Furthermore, even if we cannot know precisely all the functions of non-protein coding DNA, isn't it just possible we still haven't discovered their functions yet? Jonathan Wells in his book, *The Myth of Junk DNA*, writes there simply is no junk DNA. It is all useful.

As we have previously mentioned, evolutionists have postulated "the molecular clock." This concept originated with Linus Pauling and Emile Zuckerkandl in 1962. They suggested that by comparing DNA sequences and their protein products you could determine how organisms are related. The more closely the DNA resembles each other, the more closely organisms are related. If mutations have accumulated over time, the number of differences between the organisms could serve as a molecular clock. This could indicate the number of years that have passed since the DNA or proteins were the same, the concept being referred to as "molecular phylogeny." When programmed in a computer, this produces a "tree of life."

But even though this seems so clear to the evolutionists, there are significant problems with this "tree." First, from an evidentiary perspective, the "tree" does not fit the fossil record nor the molecular evidences given by evolutionists. When using the molecular clock, variation in ages of organisms differ greatly, in some cases by as much as 1, 000,000,000 years. Even for evolutionists, that's a long time! Perhaps even more importantly, comparisons of different "molecular trees" frequently generate different trees based solely on divergent anatomical characteristics and are often contradictory. Finally, these molecular models are based on several assumptions that cannot be proven, one being that mutations occur at constant rates and the other being the obvious, that mutations have the ability to produce evolution in the first place.

For the evolutionists, evolution must be true; therefore the clock hypothesis must be true, but, again, this goes nowhere because it saves evolution by believing in it in the first place. For Vanderbilt University scientist Antonis Rokas, a leader among biologists using molecular data to study animal phylogenetic relationships, "A complete and accurate tree of life remains an elusive goal." So despite claims of many evolutionists such as Richard Dawkins, Dr. Meyer states, "The evidence (molecular and anatomical) supporting a single, unambiguous animal tree are manifestly false." Predictions by Dr. Pauling and Dr. Zuckerkandl were simply wrong to assume the degree of similarity indicates the degree of evolutionary relatedness. Mutations do occur. To deny that would be ignoring the obvious. Also, mutations can cause variation in individual kinds. This is precisely how bacteria become resistant to antibiotics and parasites to antiparasitics. The question is: Can mutations cause a significant beneficial change in the morphology (or structure) of an organism?

There is very compelling evidence to suggest that this is not the case. Murray Eden, of MIT and Wistar, believes mutations had virtually no chance in producing new genetic information. Physicist Stanislaw Ulam has explained, the evolutionary process "seems to require many thousands, perhaps millions, of successive mutations to produce even the easiest complexity we see in life now. It appears, naively at least, that no matter how large the probability of a single mutation is, should it be even as great as one-half, you would get this probability raised to a millionth power, which is so very close to zero that the changes of such a chain seem to be practically non-existent." *The DNA itself doesn't suggest a step-by-step evolution in different animals.* But probably the biggest argument against mutations being the engine for evolution is that morphology is not solely determined by DNA! To illustrate this, proteins are comprised of three distinct structural levels: primary, secondary and tertiary or folds. If you were visualizing a protein you would see it folded upon itself in 3D so to speak, and it is this folding that gives the protein its functions. Proteins must create new folds to effect a change in the organism but slight changes in DNA do not fundamentally change the organism! Natural selection has nothing to help generate new folds. Another way to look at this is, for evolution to occur minor or slight mutations in DNA will not work, but we know that large scale change in DNA destroys the organism! Paul Nelson of the University of Chicago has concluded, "Research on animal development in macro evolution over the last thirty years – research done from within the Neo-Darwinian framework – has shown that the Neo-Darwinian explanation for the origin of new body plans is overwhelmingly likely to be false – and for reasons that Darwin himself would have understood."

Death Blow to Evolution: Epigenetics

A very important concept to understand at this point is that morphology, or structure of organisms, is not determined solely by DNA. In other words, there is simply a lot of genetic information beyond the DNA molecule. Neo-Darwinism gives primacy to the gene yet we know body plans are stagnant over time, and for this epigenetics may very well be the reason. Jonathan Wells has even said evolutionary theory may well fall apart based on discoveries of epigenetics.

"Epigenetic" basically means "beyond the gene" (*epi* meaning beyond). The field of epigenetics had its roots back in 1924 when German scientists, Hans Spemann and Hilde Mangold, demonstrated with experiments using newt embryos that something other than just DNA profoundly influenced the development of body plan. From Stephen Meyer's standpoint, "If DNA is not wholly responsible for body plan morphogenesis, then DNA sequences can mutate indefinitely and still not produce a new body plan, regardless of the amount of time and the mutational trials available to the evolutionary process. Genetic mutations are simply the wrong tool for the job at hand." To grasp this concept, Meyer uses the analogy of a construction site where builders use many materials such as lumber, wires, nails, drywall, piping, and windows. Yet these materials do not determine the floor plan of the house or the building being assembled. In the same way, DNA does not by itself direct how individual proteins are assembled into larger systems such as cell type, tissues, organs, and body plan – during animal development. Instead, the three-dimensional structure or spatial architecture of body plans is determined during embryogenesis by other sources or, in other words, epigenetic information in the cell. There are several epigenetic sources for this. For example, microtubules which make up the cytoskeleton of the cell are made by tubulin (gene products) that generate from each other, not DNA. Histones, an organelle in the eukaryotic cell, are involved in the sequencing of proteins which regulates storage and activation of DNA in cells of the eukaryote type, and are not DNA and could not have been put together in a piecemeal fashion. Furthermore, centrosomes, from the mother egg, form poles for cell division and replicate themselves but have no DNA. Also, ion channels and electromagnetic fields have significant morphologic effects and are not derived from DNA. Finally, the sugar code, consisting of sugar molecules in certain arrangements, influences embryonic development and protein patterns in the cell membrane and is transmitted directly from parent to daughter membranes.

What's the point then? The point is simply this: epigenetic structures are not vulnerable to mutations and even if they were it would result in the death of the organism. Traditional Neo-Darwin arguments do not work with epigenetics.

Of course, evolutionists propose a mechanism for epigenetics but it presumes evolution and is really no more than a story. For example, they have tried to solve the problem of epigenetics with the so-called HOX genes, genes which regulate the expression of other genes involved in animal development, but this really doesn't answer the question.

As a result of discoveries in epigenetics, evolutionary biologists themselves are now beginning to doubt Neo-Darwinian mechanisms for evolution. For example, Keith Stewart Thomson, formerly of Yale University, expressed doubt that large-scale morphological changes that accumulate over time and produce minor changes at the genetic level could ever cause whole cell changes in the organism. Geneticist George Miklos, of the Australian National University, has argued Neo-Darwinism fails to provide a mechanism that would produce large-scale innovations. Finally, Biologists Scott Gilbert, John Opitz, and Rudolf Raff have attempted to develop a new theory of evolution to explain epigenetics. They have written, "Starting in the 1970's, many biologists began questioning its (Neo-Darwinism's) adequacy in explaining evolution. Genetics might be adequate for explaining microevolution, but micro-evolutionary changes in gene frequency are not seen as able to turn a reptile into a mammal or to convert a fish into an amphibian. Microevolution looks at adaptations that concern the survival of the fittest, not the arrival of the fittest."

Going back to the HOX gene explanation, it is very doubtful that mutations in HOX genes could transform animal life. This is because HOX genes coordinate the expression of many different genes and experimentally generated mutations in HOX genes have universally proven harmful. Secondly, HOX genes are typically expressed after the beginning of animal development and well after the body plan has begun to be established, therefore, they could not be involved in determining body plan formation. Finally, HOX genes provide information for building proteins that function as switches that turn genes off and on but the HOX genes themselves do not contain information for building the structural parts. In other words, HOX genes do not have all the genetic information and HOX genes certainly do not possess epigenetic information. In fact, it is quite the opposite. Epigenetic information in structures determines the function of many HOX genes.

Finally, evolutionists have looked at developmental gene regulatory networks (DGRN) as a possible cause for the creative power of evolution. Developmental biologists have found that gene products, in the form of proteins in our RNA necessary for the development of animal plans, transmit signals that influence individual cells' development. These signaling molecules also form circuits or networks of coordinated interaction much like a circuit board. Work done by Dr. Eric Davidson at the California Institute of Technology, however, has demonstrated these so-called DGRNs actually make the organism resistant to change and cannot vary without causing catastrophic consequences. Dr. Davidson is quoted in *Darwin's Doubt* saying, "There is always an observable consequence if a DGRN's sub circuit is interrupted. Since these consequences are always catastrophically bad, flexibility is minimal, and since the sub circuits are all interconnected, the whole network partakes of the quality that there is only one way for things to work. And indeed the embryos of each species develop in only one way."

Epigenetics then represents a profound challenge to Neo-Darwinism mechanisms. Once again, the kind of mutations evolution needs do not occur, and the small changes (adaptation) cannot be the mechanism for large changes even if this seems so intuitive to the evolutionists!

Is DNA Like A Computer?

There are other aspects and functions that occur in DNA that one would not predict using an undirected evolutionary model. For example, molecular biologists have discovered that multiple messages (sets of assembly instructions) can be stored in the same sequence of bases or regions of the genome and sometimes these messages overlap along the genome. Splicing, editing, and reading processes can produce more than one protein from the same RNA message. W.Y. Chung, a bio-informaticianist at the Center of Comparative Genomics and Bioinformatics at Penn State University, "has noted the existence of 'dual-coding' and overlapping protein-coding reading frames, just one of many cellular innovations for concentrating genomic information, is 'virtually impossible by chance.'" Also, genetic messages are encoded within genetic messages, sort of like the spy encryption devices in World War II. There is a hierarchal arrangement that is optimized for access and retrieval, like a filing system with folders. All these processes would be expected from an intelligent design point of view but not evolution.

Orfan Genes

Open reading frames of unknown origin, so-called ORFAN genes, also pose a dilemma for the evolutionists. There are many hundreds of thousands of genes in many diverse organisms that do not exhibit any significant similarity in sequences to any other known genes. These genes have no homolog and cannot be explained by evolution. Some might argue that as more are mapped, homologs will become apparent, but in reality more ORFAN genes are being found with no evidence that the trend is stopping.

Neo-Darwinism and Math

We have observed before, some have used computer programs to predict evolution but in every case an intelligent programmer provided necessary "active" information when producing these models. Meyer has pointed out in *Signature in the Cell*, "Typically, genetic algorithms may lack realism 1) by providing the program with a target sequence, 2) by programming the computer to select for proximity to future function rather than actual function or 3) by selecting for changes that fail to model biologically realistic increments of functional change, increments that reflect the extreme rarity of functional sequences of nucleotide bases or amino acids and the relevant sequence space."

Here, however, are some equations that are difficult to dispute. Remember it takes three bases (or codons) to generate one of twenty protein forming amino acids. Some genes have a thousand bases (300 amino acids for an average protein). This length protein represents just one possible sequence with the probability of that sequence being 20^{300} or 10^{390}. That is a very large number, especially when you consider there are only 10^{69} atoms in the Milky Way!

Here is one more bit of interesting math. Coordinated mutations are necessary for evolution to occur. Michael Behe and David Snoke calculated that mutations in selection could possibly generate two coordinated mutations in a mere one million generations, but that would take a population of one trillion or more multicellular organisms, a number that is completely impractical and exceeds the size of affected breeding populations. Conversely, they found mutations and selection could generate two coordinated mutations in a population of one million organisms, but only if the mechanism had 10 billion generations at its disposal. Since 10 billion generations computes to 10 billion years, twice the supposed age of the earth, this represents an unreasonable length of time to wait for the emergence of only a single gene, let alone any other significant changes that could occur evolutionarily.

Behe published these works as well as other findings in his book, *The Edge of Evolution*, after which he was highly criticized by evolutionists, especially two out of Cornell University, Rick Durrett and Deena Schmidt. These evolutionists published a paper, "Waiting for Two Mutations: With Applications to Regulatory Sequence Evolution and The Limits of Darwinian Evolution." Their calculations suggested it would not take several 100 million years but "only 216 million years to generate and fix two coordinated mutations in the hominid line." Unfortunately, that is more than 30 times the amount of time available to produce humans and chimps and all their distinctive complex adaptations according to evolutionary theory. So as we can see, calculations performed by both critics and defenders of Neo-Darwinian evolution actually reinforce and come to the same conclusions; that is, if coordinated mutations are necessary to generate new genes in proteins, Neo-Darwinian math establishes the implausibility of the mechanism. It is no wonder that evolutionists now are turning away from Neo-Darwinism and looking for other mechanisms for evolution.

Cytochrome C
Evolutionists have also used the similarity of the cytochrome C protein in apes and men as an argument for evolution. The story of cytochrome C

(a protein involved with energy distribution in the cell), however, is really a profound argument against it! Let's look at why.

The cytochrome C molecule has 100 amino acids and forms a 3D configuration. The amino acid sequence varies from organism to organism but is found in bacteria, fungi, high plants, and vertebrates. The Dayhoff *Atlas of Protein Sequence and Structure* categorizes 1,089 entries in a matrix form for cytochrome C. As organisms increase in distance from one another their amino acid sequences vary more and more. What is key is there are *NO* transitional or intermediate classes. They are absent from the matrix! This may be a bit technical but all organisms from humans to yeast exhibit sequence divergence by 64 to 67% from the bacteria cytochrome. This means, once again, there are no intermediates between bacteria and eukaryotic cells. Michael Denton has written, "This is almost mathematical precision, the isolation of fundamental classes of organisms at the molecular level." Denton also states, "There is not a trace at the molecular level of the traditional evolutionary series: Cyclostome à fish, à amphibian à reptile, à mammal. Incredibly, man is as close to the lamprey as are fish." Thus there is incredible orderliness of all divisions biochemically, a phenomenon that fits perfectly if there was an intelligent designer!

Predictions
The validity of a scientific theory rests a lot on its ability to predict. Creationism and intelligent design have been ridiculed for the apparent inability of these concepts to predict. In a debate between evolutionist Bill Nye and creationist Ken Ham in 2014, this very idea was put forth by Mr. Nye. He claimed, erroneously, that evolution makes predictions and that creationism and intelligent design do not. Although in that debate some predictions were offered by Mr. Ham, I thought the discourse on the subject was a bit lacking. The truth is there are quite a few predictions that intelligent design can make, many that would not be predicted by naturalism. In Stephen Meyer's book, *Signature in the Cell*, he lists at least twelve predictions that intelligent design has made. I would like to just summarize a few of these at this point. One prediction is that the fossil record will show a top-down phenomenon and not a bottom-up phenomenon which has in fact been verified. Another prediction involves the junk DNA we have already discussed. Creationists predicted "junk DNA" would be found to have functionality, and more and more this is being found true. Intelligent design and creationism also predict that organisms have limits to change, meaning there are limits to the variability of the various kinds. Observationally, this has precisely been shown.

Another intriguing prediction of intelligent design concerns the science behind finding a cure for cancer. Scientists in the past have primarily looked at the DNA and mutations that occur in DNA when looking for the cause of cancer. Jonathan Wells, on the other hand, reasoning from a design point of view, postulated the centrosome (a cellular organelle) may be the culprit in some cases. This is turning out to be likely and more and more scientists are beginning to look at this. Finally, Michael Behe, reasoning from the irreducible complexity point of view, a concept we will discuss later, predicts more and more mini-machines within the cell will be discovered that carry on more functions than we currently realize. These mini-machines such as the flagella mentioned in Behe's book, *Darwin's Black Box*, demonstrate incredible design, design that would most assuredly not have occurred by chance. So the point here is that to criticize creationists and the intelligent design community for not being able to make scientific predictions based on their concepts is patently wrong!

Of Apes and Men

We have already seen the argument that similarities between apes or chimps and man morphologically and genetically are given as proof of common ancestry and therefore, proof for evolution. That there are similarities is not really the question, and if an organism is, from a homological point of view, very similar, would it not stand to reason they would be genetically similar as well? But how similar are apes and men and is evolution an adequate explanation? Before the completion of the Human Genome Project, Mary-Claire King of the University of California San Francisco and Allen Wilson of the University of California at Berkley in 1975 argued there was only a 1% difference between chimps' and humans' DNA. This was somewhat surprising even to them because this was less than similarity for morphology would even suggest.

The human genome was mapped out in 2003 and the chimpanzee genome in 2005. In 2005, Pascal Gagneux reported researchers were finding out that the 1% distinction might not be accurate. Chunks of missing DNA, extra genes, and altered connections confound any qualification of "humanness" versus "chimpness" changing this 1% to more like 3% difference. With the differences in the numbers of copies of DNA this number becomes 6.4%, the greatest differences between chimps and humans as one may or may not expect having to do with genes that code for brain functions. Furthermore, the genome for the Y chromosome in chimpanzees was not completed until 2010. It was discovered that the human chromosome had 78 genes in the

Y chromosome and the chimp only 37, a 58% difference not explained by traditional evolutionary process but only by some "rapid evolution." Keep in mind, even a 1.23% difference in genes represents 35 million differences! With substitutions and insertions this becomes 40 million differences. To put this in perspective, an 8 ½ x 11 page can hold 4,000 letters. It would take 10,000 pages to equal 40 million letters. Is this a lot? I guess it is in the eyes of the beholder!

Further differences involve telomeres (repeating sequences at the end of a sequence). Chimps have 23,000 and humans have 10,000 base pairs in these telomeres. Also, humans have 689 genes chimps lack and chimps have 86 genes that humans lack. It is important also to keep in mind a small difference in gene sequences can mean a large difference in function. An example of this is the Fox P2 protein involved in language. Only two out of seven amino acids differ between chimps and humans, 99.7% identical, but has anyone seen a chimp talking? Also, as we have seen, humans have 23 pairs of chromosomes and chimps 24. Evolutionists say two of the chimps' chromosomes were fused together before man evolved and somewhat deceitfully will label chromosomes 1 and 2 in chimpanzees chromosome 1a and 1b. They do not, however, offer any evolutionary cause for this fusion or give a natural selection process for it. Finally, evolutionists also point to the similarity in junk DNA shared by chimps and humans but here again as we have seen there really is no such thing as "junk DNA."

One of the biggest hurdles for evolutionists to jump in regard to the argument for common descent of apes and men has been referred to as "Haldane's Dilemma." There are, per evolutionists, 40 million total separate mutation events that separate humans and chimps. Haldane calculated an average mutation rate of 10^6 with a population size of one million, a number compatible with evolutionists. Given evolutionary timescales, this implies 150,000,000,000 forerunners to modern men! The obvious question is, where are the fossils? There should be innumerable fossils if humans had this many forerunners. Furthermore, population studies would also make this number utterly impossible. There cannot be and there are not thousands of millions of "pre humans."

How can evolutionists explain this? They cannot without referring to abductive reasoning. Abductive reasoning is illustrated this way: If P then Q, Q therefore P or in other words, if humans and chimps share a common ancestor there will be evidence of chromosome similarity. There is chromosome similarity, therefore humans and chimps share a common ancestor!

What can we say then, are there similarities between humans' DNA and chimps'? There certainly are, but there are also significant differences. But what explains these differences more accurately, design or evolution? Even from an evolutionary point of view there simply isn't enough time for mutations to have produced humans from chimps even if mutations were capable of doing so, which we have seen they are not. Overwhelmingly, design by a common designer is the explanation for the similarities between chimps and humans. Wouldn't you agree?

Summary

DNA, genetics, and the genome offer no more proof for evolution than homology did, in the past, or the fossil record for that matter. In fact, it is very presumptuous to say evolution has occurred when it is impossible to determine how even a single functioning protein could have occurred naturally! Francis Collins, the head of the Human Genome Project, understands this and yet still clings to evolutionary theory. Dr. Collins, in his support for modern Darwinism, seems unaware of the arguments we have just presented. Stephen Meyer points out in his book, *Darwin's Doubt*, neo-Darwinian mechanism fails because: (1) "It has no means of efficiently searching combinatorial sequence space for functional genes and proteins and, consequently, (2) it requires unrealistically long waiting times to generate even a single new gene or protein. It has also shown that the mechanism cannot produce new body plans because: (3) early acting mutations, the only kind capable of generating large-scale changes, are also invariably deleterious, and (4) genetic mutations cannot, in any case, generate epigenetic information necessary to build body plan."

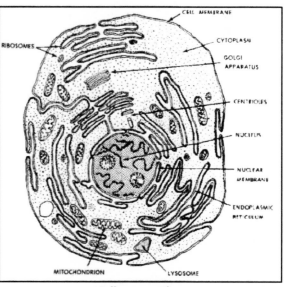

Cell cross section

We have seen that in all organisms the roles of DNA, mRNA, and proteins are identical and no living system can be thought of as "primitive." Even the bacteria cell is an intricate machine more complicated than the greatest of human inventions. These similarities do not speak to evolution but to design. Magnify a cell a thousand million times until it is 20 km in diameter and it will resemble a giant ship, large enough to cover a city the size of New York.

Michael Denton has stated, "It is the sheer universality of perfection, the fact that everywhere we look, to whatever depth we look, we find an elegance and ingenuity of an absolutely transcending quality, which so mitigates against the idea of chance." I believe this observation by Stephen Meyers concludes the case for intelligent design nicely, "The case for design restores Western thought to the possibility that human life in particular may have a purpose or significance beyond temporary material utility." Yes indeed! As Ken Ham responded to Bill Nye in their debate, "It's in the BOOK."

Chapter 7

Irreducible Complexity

What should be becoming clearer and clearer as we study Darwin's theory is that it cannot account for the molecular structures of life nor can it explain the origins of life. Even today, evolutionary literature remains relatively silent regarding how molecules produced life.

In the early 19th century, Matthias Schleiden and Theodor Schwann put forth the cellular theory of life which over the next 200 years has largely been verified; that is, all life is made up of either singular cells or a collection of many cells. Each one of us began as one cell that eventually divided over and over to become trillions of cells, yet we function as one unit. Each cell is made up of a nucleus with many other cellular organelles encased in a membranous structure. These organelles are given their structure and function through DNA programming. Nineteenth century scientists, though they knew about structures in the cell, had virtually no understanding of their biochemical functions. For Darwin, the cell was a "black box"; that is, a thing of intense mystery, much like a computer would be a "black box" for myself and many of you.

In 1996, Michael Behe published the now classic *Darwin's Black Box*. This seminal work, a must-read for anyone serious about understanding the creation vs. evolution debate, introduced the concept of irreducible complexity. Modern science now has a much better understanding of the biochemistry of life and can better describe precisely how vision occurs, how blood clots, how the body fights off infections, and many, many more vital functions that define life. These are very complex biochemical systems that are still being

Michael Behe

studied. The question Behe asked, and rightly so, was how these systems could be gradually produced, which is a prerequisite for evolution. You see, it is one thing to propose a mechanism for the gradual changes that must occur morphologically or structurally for an organism to change, and quite another to describe biochemical gradualism which is even more necessary.

Darwin's theory demands gradualism. Darwin stated, "If it could be demonstrated that any complex organ existed which could not possibly have been formed by numerous successive, slight modifications, my theory would absolutely break down." Are there systems that could not be formed by "numerous, successive, slight modifications"? It turns out there are many. Behe refers to these systems as irreducibly complex. This concept is really not that difficult to understand and evolutionists, though they have tried, have not been successful in refuting it.

What is irreducible complexity? Let's see what Dr. Behe means by it. "By irreducibly complex I mean a single system of several well-matched, interacting parts that contribute to the basic function, wherein the removal of any one of the parts causes the system to effectively cease functioning." The system cannot just appear as it requires successive modifications of precursors to function and these precursors would, by evolutionary dogma, require some function of their own. In other words, if a biologic system cannot be produced gradually it would have to arise as an integrated unit, in one fell swoop, for natural selection to have anything to act on. A mutation, the motor for evolution, cannot change all the instructions in one step.

Darwin proposed numerous small changes must occur anatomically for a change in the organism to occur, like the development of the human eye for example. He even proposed these small changes for the eye in his book, *On the Origin of Species*. While this may have some logic (though certainly not proven), when investigating vision on a chemical level, a thing we can now do to a large extent, Darwin's theory fails on an even grander scale. We will discuss some of these biologic systems subsequently but first let's define irreducible complexity a little bit more. For this, Behe used the analogy of a mouse trap.

A mouse trap, though very simple, represents an irreducibly complex system. As we all know a mouse trap functions to immobilize a mouse. It has a platform, a holding bar, a hammer, a spring, and a catch. For the mouse trap to function, all the mechanisms must be present and functional with the loss of even one of them causing the mechanism to fail. None of the components has a function in catching a mouse by itself. If you were "evolving" a mouse trap, how and why would any of these components develop if they had no function? These components would have to have some minimal function to exist. Natural selection demands at least minimal function and minimal function is critical in

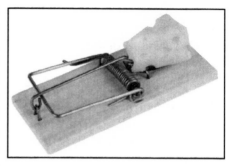

evolutionary processes. Behe believes, "irreducibly complex systems are nasty road blocks for Darwinian Evolution, the need for minimal function greatly exacerbates the dilemma."

There are many irreducible systems in all organisms including, of course, mankind. Behe alludes to several of these in his book, *Darwin's Black Box*, including cilia, flagella, the antigen-antibody and complement cascade, the Bombardier beetle, vision, and the blood-clotting cascade. We cannot review all of these; but I would like to describe vision and the clotting cascade, as well as one system I find fascinating, the endocrine system of the human body, which I believe certainly fits the criteria of being irreducibly complex. The point in discussing these systems is not to rack our brains but to realize just how complex they are, and how utterly ridiculous evolutionists appear when trying to explain them. When examining irreducibly complex systems, it might be helpful to think of Rube Goldberg machines as Behe does. We have all seen pictures of Rube Goldberg's machines. These are cartoons and seem somewhat silly but when you follow them precisely to the end results, they still work. These cartoons are for entertainment and

Rube Goldberg machines

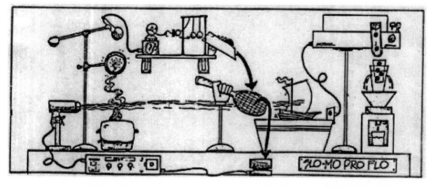

always get a big laugh but they are nonetheless irreducibly complex. Modern biochemists have discovered many Rube Goldberg-like systems. Let's now look at just a few of them.

Vision

Nineteenth century anatomists knew the anatomy of the eye very well. The pupil, which acts as a shutter, the lens which gathers light and focuses it, the retina which forms the sharp image, and the muscles of the eye which allow quick movement that all make up the modern mammalian eye were certainly known by 19th century scientists. Darwin, as we have seen, knew the anatomy of the eye well. Though Darwin attempted to explain how an eye could have evolved in a step-by-step fashion, he did not try to explain where the starting point might have been and he certainly could not explain how a nerve could have become sensitive to light in the first place.

Modern man still can't explain everything about vision biochemically but we have at least begun to approach an understanding of it. What follows is our best explanation to date of how vision works. When light hits the retina, a photon interacts with the molecule 11-cis-retinal, which rearranges instantly (in picoseconds) to form a completely different molecule called trans-retinal. This change in trans-retinal forces a change in the shape of the protein called rhodopsin, which then metamorphosizes into metarhodopsin II, sticks to a protein called transducin before said transducin had been tightly bound to the molecule GDP. GDP then falls off and GTP now binds with this transducin. The GTP-transducin-metarhodopsin now binds with the protein phosphodiesterase in the membrane of the cells. With this attachment phosphodiesterase "cuts" a molecule called cGMP. Phosphodiesterase lowers cGMP's concentration.

Then when cGMP is reduced an ion channel closes causing positively charged sodium ions to be reduced. This "imbalance in charge" causes a current to be transmitted down the optic nerve to the brain, resulting in vision.

For supplies of 11-cis-retinal, cGMP and sodium ions not to run out, something has to turn off the proteins which turn them on to begin with. First, dark ion channels let calcium into the cell, calcium is pumped back out, cGMP levels off, shutting down the ion channels, and calcium goes down too. Second, a protein called guanylate cyclase resynthesizes cGMP. Third, at the same time metarhodopsin is modified by rhodopsin kinase, then binds to a protein, arrestin, which prevents rhodopsin from activating transducin.

Trans-retinal eventually falls off of rhodopsin, converts back to 11-cis-retinal, and binds to rhodopsin to get back to the starting point for another cycle. This then is the simplified version of the biochemical explanation of vision, which as you can see is a very complex mechanism. The question then is: How did all these chemicals come into being and what function did they have before vision was accomplished and what would be the purpose of said function? This is irreducible complexity.

No longer then is it enough to explain anatomical changes evolutionarily. As Behe points out, "Anatomy is quite simply irrelevant to the question of whether evolution could take place on the molecular level." On the contrary, structure becomes irrelevant when trying to explain evolution, and for that matter so does the fossil record.

The Clotting Cascade

Punch a hole in a milk carton full of milk and all the milk above the hole will run out. Punch a hole in your finger and unless you have a serious bleeding disorder all the blood will not run out of you and you will not die. Clotting is a very complex system producing homeostasis, which is critical for life. If you clot too slowly you can bleed to death from the most trivial of injuries, yet if you clot too fast a heart attack or stroke can ensue. Scientists have a very good understanding of the clotting mechanism, and as a result there are many medications that act on various areas of the clotting cascade to both increase the mechanism as well as slow it down in order to protect from strokes and myocardial infarctions. Let's look at this irreducibly complex system.

For clotting to occur, fibrinogen must be present in the plasma. Fibrinogen constitutes about 2 to 3% of the plasma and looks like a set of barbells with weights in the middle. When a cut occurs, thrombin slices some of the fibrinogen chains and fibrin sticks together to form a patch, hence a clot forms. What would happen if thrombin continued to work without a controlling feedback mechanism? The entire arterial system would form one massive clot; but fortunately for us this is not what happens. Our bodies must and do control the activity of thrombin.

The clotting mechanism has been referred to as a cascade. For homeostasis to occur, every part of the cascade must work perfectly and removal of any part of that cascade could result to the detriment of the organism.

Thrombin initially exists as prothrombin. Prothrombin must be activated by the Stuart factor. Stuart factor cleaves thrombin to activate thrombin to cleave fibrinogen for fibrin to form the clot. Accelerin is needed to increase the activity of the Stuart Factor. Accelerin exists in the inactive form (pro-accelerin) which in turn is activated by thrombin (there is always some bit of thrombin in the bloodstream).

Backing up, prothrombin, before it can be turned into thrombin, must be modified by glutamate residues changed to gamma carboxyglutamate. For this to occur, Vitamin K must be present. In its presence, prothrombin cannot be active, which slows the clotting mechanism down. Stuart factor is activated by intrinsic and extrinsic pathways. With the intrinsic pathway, all the proteins for clotting are in the plasma. In the extrinsic pathway, some clotting proteins occur on the cells.

Let's look at the intrinsic pathway. An animal receives a cut and as a result a protein called Hageman factor sticks to the cells' surface near the wound. Bound Hageman factor is cleaved by the HMK protein and yields activated Hageman. Hageman converts prekallikrein to kallikrein, which speeds up the conversion of more activated Hageman. Hageman and HMK transform PTA to activated PTA, which activates convertin which activates the Christmas factor. Christmas factor with antihemophilic factor changes Stuart factor to its active form.

With the extrinsic pathway, proconvertin becomes convertin (by acti-vated Hageman and thrombin with tissue factor). Convertin changes Stuart factor to its active form. Tissue factor appears only outside the cell and is usually not in contact with the blood. Only when injury brings tissue factor into contact with the blood is the intrinsic pathway initiated. Intrinsic and extrinsic pathways then cross.

It is obviously key that the clotting cascade has to be turned off. Also keep in mind that in the clotting cascade, none of the components of the system are used for anything but controlling the function of the clot, with the exception of calcium which is also part of it.

What stops the clot? Antithrombin binds active clotting proteins, these bind to heparin (found inside cells of undamaged blood vessels). A protein known as protein C then destroys accelerin and activates antihemophilic factor. Thrombomodulin then binds with thrombin and the clot is stabilized by fibrin stabilizing factor. Eventually the clot is removed by plasmin (acti-vated from its inactive form plasminogen by a protein called tPA). Another protein, alpha-antiplasmin, also controls clot dissolution by binding to the fibrin clot.

In addition to our discussion there are yet two other ways the body also prevents bleeding. One is contraction of blood vessels and the other is platelets. When vessels vasoconstrict bleeding is impeded, and platelets help to form yet again a sticky patch to prevent blood loss.

I know this is somewhat complex and detailed, but once again the point is to understand that all components are necessary for clotting and hemostasis to occur. People who lack some of these various clotting factors are hemophiliacs and bleed easily, to their significant detriment. For example, the second most common hemophiliac disorder is the lack of the Christmas factor. To help see the clotting cascade, I have provided a diagram.

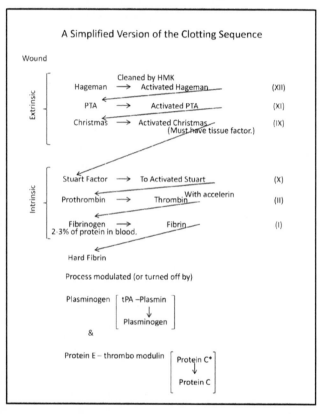

Have evolutionists tried to explain the clotting mechanism? It turns out they actually have. Michael Behe quotes Russell Doolittle, Professor of Biochemistry at the University of California - San Diego extensively. Dr. Doolittle has tried to put an evolutionary explanation to the clotting cascade and though we won't go into that in detail here, I would like to point out that Professor Doolittle's explanation is contrived and uses a yin and yang scenario. He offers no causation for the various clotting factors and frequently uses words such as "appear," "spring forth," "are born," "arise," and are "unleashed" to describe where the clotting factors came from. These terms are at best nonscientific and explain absolutely nothing. But the most serious flaw in Dr. Doolittle's evolutionary explanation of the clotting cascade concerns irreducible complexity. In Doolittle's account, clotting does not occur until the last step. This begs the question as to how to get from one step to the other without the organism bleeding to death first! If this is the case, what purpose would the previous steps serve for the millions of years it was evolving until the last step was added and the anti-clotting mechanism became functional? Behe points out, "Darwin's mechanism of natural selection would actually hinder the formation of irreducibly complex systems such as the clotting cascade." Doolittle's "explanation" is nothing more than just a story. "The fact is, no one on Earth has the vaguest idea how the coagulation cascade came to be," so says Dr. Behe.

The Endocrine System

The last complex system that defies an evolutionary explanation that I would like to examine concerns the endocrine system, specifically the pituitary gland and its regulatory function. The pituitary gland, located at the base of the brain and protected in a bony structure called the sella turcica, has been called the master gland of the body. The pituitary gland produces eight hormones that regulate homeostasis, with itself being up regulated by the hypothalamus. Hormones are signaling substances produced by glands transported by the circulatory system to target distant organs to regulate physiology or behavior. Hormones secreted by the pituitary gland help in controlling growth, blood pressure, some aspects of pregnancy, aid in the production of breast milk, control sex organ functions, metabolism, temperature, water osmolarity, and thyroid functions. Each one of these functions is produced at a distant site from the pituitary. For example, the thyroid gland is a gland that resides in the neck on either side of the trachea. All of these endocrine glands form a complex feedback mechanism through the pituitary to control all essential bodily functions. We cannot look at all of these, but I would like to just briefly see how the pituitary gland controls thyroid function.

The thyroid gland produces thyroid hormones which help control the metabolism of our bodies. Too much thyroid hormone results in a disease known as Grave's disease. Symptoms of Grave's disease include weight loss, excessive sweating, fast heart rate, and diarrhea – all caused by hyper-metabolism. Decreased thyroid function results in hypothyroidism, many times in the form of a disease known as Hashimoto's thyroiditis. Hypothyroidism produces weight gain, cold intolerance, constipation and other symptoms that are a consequence of low metabolism. To function properly, the thyroid gland must receive information from the pituitary which, as stated earlier, receives information from the hypothalamus. Thyroid regulating hormones from the hypothalamus interact with the pituitary gland to form thyroid stimulating hormones, which further interact with the thyroid gland to produce more thyroid. For example, if the thyroid is functioning normally, thyroid-stimulating hormone, which can be measured with a blood test, will be in a normal state. If for some reason the thyroid gland underproduces, thyroid-stimulating hormone will increase. When the thyroid gland is overproducing, thyroid-stimulating hormone will in turn decrease in an attempt to slow thyroid production down. This feedback system is critical to maintain proper metabolism and homeostasis; the failure in any part could lead to serious disease. From my standpoint as a physician, this is an example of irreducible complexity and it is hard to imagine how this system could have evolved. A system is useless and can serve no useful function until it is "fully integrated" when the last piece is finally added after supposedly millions of years. All these systems mentioned, as well as many others, serve no useful purpose unless completely integrated and activated.

Final Thoughts

Evolution has no answer to irreducible complexity. Some, including Dawkins, have tried to discredit irreducible complexity by claiming the individual components of a complex system may have a function independent of the final system, but they offer no precise examples. In the end, they are only *ad hoc* explanations. What's amazing to me is how evolutionists try to make Darwinism appear very simplistic, as if it is just a breeze! The problem is, for evolution to occur it's not a breeze, and as we have seen repeatedly it requires a much larger leap of faith than creationism.

Biochemistry was not a part of evolutionary synthesis, yet Darwin's theory must account for molecular structures in life and it simply does not! As a result, you have even more and more silliness propped up by such nonsense as Richard Goldschmidt's and O. H. Schindewolf's "Hopeful Monster" from the 1940's and Stephen J. Gould's and Niles Eldredge's "Punctuated Equilibrium." This has caused Lynn Margulis, the late distinguished university Professor of Biology at the University of Massachusetts, Amherst, to predict history will judge NeoDarwinism "as a minor 20th Century religious sect within the sprawling religious persuasion of Anglo-Saxon biology." Dr. Margulis has challenged evolutionists to give just one example of the unambiguous formation of a single new species formed by the accumulation of mutations, and to date her challenge has gone unmet.

Chapter 8

Creating Life

The idea that life came forth from a "little pond" or "pre-biotic soup" has been around since Darwin first proposed it in 1859, and now has permeated our collective conscious to the point of becoming iconic. The 1940 Disney cartoon movie *Fantasia*, depicted a primitive form of life crawling out of the ocean to solidify the icon even more, only now into children's minds! One would think if life could spring forth undirected, by all naturalistic causes, it should be fairly simple for scientists to pre-assemble the necessary components and conditions to create some rudimentary form of life. Though many have attempted this, none has been successful.

Stanley Miller, in the 1950's, using a supposed prebiotic soup created amino acids, the building blocks of proteins. His work was heralded as one of the greatest achievements in the history of mankind. Science writer, William Day, wrote at the time about the Miller experiment, "It was an experiment that broke the log jam. The simplicity of the experiment, the high-yields of the products, and the specialized biologic compounds produced by the reactions were enough to show the first step in the origin of life was not a chaotic event, but was inevitable."

Then in 1967, RNA was supposedly created in the lab and by the 1980's, scientists Thomas Cech and Sidney Altman demonstrated that RNA can sometimes act like an enzyme. With a giant leap of faith, biologists such as Walter Gilbert postulated RNA may be able to be synthesized without proteins, thus proposing an RNA world. More recently, in 2010, J. Craig Venter's team, using a computer, synthesized the largest piece of DNA ever, and placed it in a bacterium cell thus producing the first human-designed life form. Accolades for this achievement were especially grandiose. Some called it "the most important achievement in the history of mankind" with the *Financial Times* stating, "It seems to extinguish the argument that life requires a special force or power to exist." Andrew Brown of the London paper *The Guardian* called it "a complete victory for materialism" and even Venter himself, referring to this new man-made self-replicating species, called it "a philosophical advance as much as a technical one."

The common thread running through these experiments is the attempt by scientists to create life or at least the building blocks for life. For our discussion, we shall first consider biologic life not artificial intelligence, a topic to be discussed later. We will not discuss cloning, which of course is not really creating new life at all, but instead reproducing life asexually.

The question then is: Has mankind really created life *ex nihilo* or from scratch, and do these experiments confirm evolution? To answer this we will need to look at each of these achievements and investigate exactly what they demonstrated. Since the Miller experiment is the oldest and most iconic, we will discuss it first.

The Miller Experiment

The Miller Experiment, also referred to as the Miller-Urey Experiment, was an attempt in the 1950's to produce the building blocks, amino acids, from a supposed prebiotic soup. Stanley Miller was the graduate student of advisor Harold Urey. In the 1920's, two scientists, A.I. Oparin, a Russian, and J.B.S. Haldane, British, independently suggested that lightning striking some

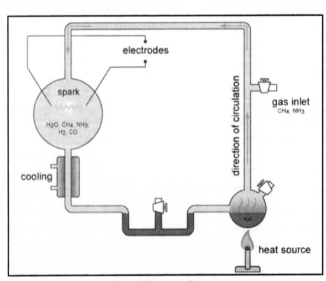

The Miller Experiment

chemical mixture in the primitive atmosphere could produce amino acids, which in turn could eventually lead to life. Miller, in 1953, built an apparatus, placed chemicals in it he thought were present at the dawn of life with an atmosphere conducive for life, and ran an electric current through it, producing a spark which eventually led to the production of amino acids. Miller's work was lauded as a stupendous success and is still featured in most high school biology textbooks; yet most geochemists now realize the experiment failed and had nothing to do with the origins of life. Why did they reach such a conclusion? Let's read on.

First, to understand why the Miller experiment failed we need to look at the earth's atmosphere and what scientists believed it to be before life began. Presently, the earth's atmosphere contains 21% oxygen, which is great and even necessary for life to exist now, but paradoxically would have prevented life from forming initially from scratch. Oxygen is used for aerobic respiration which in turn breaks down organic molecules, oxidizing them in other words. Synthesis is the process of building up and needs a reducing atmosphere. As strange as it intuitively seems, too much oxygen then would prevent the beginning of life according to evolutionist. It is fatal for synthesis. Some compartments in living cells actually exclude oxygen from the synthesis process. This is partly why antioxidants are so popular today; we are trying to prevent cellular aging. For the origin of life, the chemical building blocks could only have been formed in a primarily non-oxygen atmosphere. Oparin and Haldane, Miller's predecessors, realized this, and proposed the primitive atmosphere was strongly reducing and high in hydrogen, which is the opposite of what we have today. They had no proof of this, but realized that it would be the only way amino acids could ever occur "naturally." Lightning then could produce the energy necessary for this to occur.

In 1952, Harold Urey (Nobel Prize winner in chemistry) concluded that the atmosphere of early earth was composed of hydrogen, ammonia, methane and water vapor. At the time, scientists postulated the earth was created by interstellar dust mostly composed of hydrogen. With this information in hand, Stanley Miller at the University of Chicago, a Ph.D. candidate of Urey's, assembled a glass apparatus, pumped oxygen out of it (a good thing or Mr. Miller might not have survived his own experiment) and replaced it with water, methane, ammonia, and hydrogen, circulated the mixture through high voltage, and presto, in one week styrene and alanine (amino acids) and a sludge of mostly reactive products not found in living organisms resulted. For this work, Miller needed a "cold trap" or his chemical products would have been destroyed. Per Richard Blass, Gary E. Parker, and Duane Gish, "It is important for us to realize that the energy which forms molecules is also the energy that destroys these molecules as they are formed."

The biggest problem for Miller's experiment is evidence from scientists that now believe the early earth atmosphere was not what Urey, Oparin and Haldane had predicted. Princeton geochemist, Heinrich Holland, and Carnegie Institute geophysicist, Philip Abelson, have concluded Earth's atmosphere was not formed from interstellar gas but rather from volcanic action, which produces CO_2, nitrogen, water vapor, and traces of hydrogen. Hydrogen is difficult to keep in Earth's atmosphere because it is too light for gravity to have an effect on it. This would leave oxygen behind through a process known as photo dissociation and though scientists do not know how much oxygen was in Earth's early atmosphere, it was probably never oxygen-free, which is a big problem for the Miller Experiment. The only reason to believe it was oxygen-free is because evolutionists need a reducing atmosphere!

Two British geologists, Harry Clemmey and Nick Badham, looked at rocks supposedly dated 3.7 billion years and proved they contained oxygen, and more and more evidence continues to mount to the point now that scientists no longer believe Earth's early atmosphere was reducing, i.e. anaoerobic or lacking oxygen, and it never looked like Miller's. Marcel Florkin has determined the concept of a reducing primitive atmosphere must be abandoned; hence, the Miller Experiment is no longer believed to be adequate to explain the origin of life.

There are other problems with the Miller Experiment as well. Ammonia for example absorbs ultraviolet light from the sun; therefore, it would be destroyed by it. Also, there was no allowing for the removal of hydrogen in Miller's apparatus, yet it would have to have been lost in the Earth's early atmosphere. As a result, without ammonia or methane and only water, carbon dioxide and nitrogen, no amino acids are formed at all. Clearly, the Miller Experiment does not work in real atmospheric conditions.

In spite of all this, the Miller Experiment continues to be used in high school textbooks to explain how life could have originated in the early formation of the earth. Kenneth Miller of Brown University and Joseph Levine, science writer and producer from Concord, Massachusetts, in their high school textbook, *Biology,* described the Miller/Urey experiment, calling the results of it "spectacular." "Miller and Urey's experiment suggested how mixtures of the organic compounds necessary for life could have arisen from simpler compounds present in a primitive Earth." Clearly, Miller and Levine know the Miller Experiment has been discredited and even state in their textbook, "Scientists now know that Miller and Urey's original simulations of Earth's early atmosphere were not accurate." They still, however, consider it important enough to place in their textbook, which is extremely misleading at best. It's almost as if the facts don't matter. Why Miller and Levine as well as other biology textbook authors continue to display the Miller/Urey experiment as evidence of evolution, one can only speculate. In the case of Miller and Levine's textbook they seem to cover it up by stating that other experiments based on more current knowledge of earth's early atmosphere have produced organic compounds. In fact, one of Stanley Miller's subsequent experiments in 1995 produced cytosine and uracil, two of the bases found in RNA. It should be noted that producing other organic compounds and nitrogen bases are not the same as producing amino acids. Finally, it should be said that any origin of life experiment such as Stanley Miller's must be carefully contrived to have any hope of building compounds that produce life, otherwise all you are going to have is chemically insoluble sludge. There are just too many chemically distinctive processes working together at the same time to allow amino acids producing in a non-reducing world.

Was There Ever an RNA World?

Since proteins could not have been produced in the earth's early atmosphere, scientists have turned to RNA as the precursor to life, realizing as we have seen earlier that DNA is simply too complex and in need of too many complex amino acids to have occurred spontaneously. In 1967 RNA was allegedly produced in the lab, although this has never been replicated. We will address this subsequently; but with this information in hand, in 1987, experimenting with one kind of RNA, scientists demonstrated that it could act as a catalyst or enzyme and, therefore, function like a protein, a property Hugh Ross has said requires a "leap of faith." Since then, Harvard biophysicist Kenneth Miller has noted "nearly two decades of experiments showing that very simple RNA sequences can serve as a biologic catalyst and even self-replicate." Since RNA might be able to be synthesized without protein, this seems to bypass the problem of an oxidizing atmosphere, which has led scientists such as Walter Gilbert to propose an RNA world from whence life began.

There is a problem though. Years since the RNA world was proposed, "No nucleotides of any kind have been reported as products of spark-discharge experiments or in studies of meteorites," says chemist Robert Shapiro. How about the 1967 experiment where RNA was produced? Though textbooks still refer to this "ambiguous paper," Shapiro has traced all references to this paper on RNA synthesis. Shapiro has thus concluded it is a myth and no one else has ever replicated it. RNA simply could have never been produced in the supposed prebiotic soup. Even the concept of a prebiotic soup is a myth. Gerald Joyce of Scripps Research Institute has noted how unlikely it is that RNA could have occurred spontaneously, and even if it could it would not have survived long enough in earth's atmosphere to make any difference. At most, RNA could survive up to twelve years but millions of years are needed for RNA to work as the precursor of life. Even then, if it did survive, you are still faced with the problem of information needed to produce proteins. Where would that information come from? No one has ever shown RNA could be formed before living cells. It is clear to everyone but evolutionary scientists that life did not start with RNA.

One more comment needs to be made about RNA. There is a sort of catch-22 involved in the production of RNA as well. The presence of nitrogen-rich chemicals is needed to produce nucleotides, but the nitrogen-rich chemicals prevent production of ribose; yet both ribose and nucleotides are needed.

A Victory for Materialism?

In June of 2010, *Scientific American* listed twelve events that will change everything. One of these events was the recently published work of Dr. Craig Venter, whose team of scientists developed a bacterium with DNA sequenced entirely by a computer thus producing the first human-designed life form. Dr. Venter dubbed this, "the first self-replicating species whose parent was a computer." Venter went on to call this a philosophical advance as well. As we have noted, the *Financial Times* called this the last argument for the need of a special power for life to exist. But if this is so, what did Dr. Venter's team actually accomplish and how did they do it?

Dr. Venter has been called "the world's greatest scientific provocateur." A medical doctor, scientist, billionaire, yachtsman, Vietnam vet, and explorer, Dr. Venter is above all a showman, not one to shun the limelight. Assembling a team of scientists and spending in excess of $40 million, they created this "new species" of bacteria.

In an interview with CNN, Dr. Venter explained in short terms how this was accomplished. The process was built from "four bottles of chemicals." These chemicals were actually millions of base pairs (chromosomes). Using a computer they "assembled" these nucleotides together, transferred them to a recipient cell (yeast) and then transferred it to pre-existing bacteria.

Once inside the bacteria, the new "DNA" started producing new proteins and "totally transformed that cell into a new species coded by synthetic chromosomes."

Clearly, this was a monumental achievement and surely has scientific implications for the future, both for the good and the bad. Imagine for example, creating a lethal bacterium with potential of destroying millions of lives! But did Venter's team really create life?

It is self-evident this was not creating life *ex nihilo*. Pre-existing bacteria were used to remove chromosomes and replace them with new "synthetic ones," which were in turn produced by pre-existing base pairs. The material used was not therefore material made by human beings. Dr. Venter himself acknowledged this when asked, "Did you create life?" by CNN. His answer was, "We created a new cell. It's alive but we didn't create life from scratch." Remarkable achievement that it was, Venter's team did not extinguish the argument that life needs a special force or power to exist. To the contrary, they actually reinforced it.

Artificial Intelligence

One might ask, how about artificial intelligence, isn't that life? Artificial intelligence, a term first coined in 1956 by Noam Chomsky to describe the implementation of the essential features of computers, may have originally had its origins in science fiction. Some of the first writers to deal with this subject were Earl and Otto Bender in 1939 with the book, *I Robot*, and later in 1950 by Isaac Asimov in his short story of the same name. Since then many books, movies, and television shows capture our imagination with this theme. TV shows like *The Outer Limits* and *Star Trek*, and a myriad of movies like *2001: A Space Odyssey*, *The Matrix*, *Transcendence*, *A. I.*, *Blade Runner*, *WarGames*, and most recently *Chappi*, have all had as the central theme, artificial life. Most of these stories revolve around the idea that mankind has made machines capable of sentience and consciousness, with the consequences being that these machines revolted against their makers.

Artificial intelligence is an integral part of our society now and is everywhere, from your X-Box, to automobiles that can drive themselves. AI, as it is commonly abbreviated, has the ability to analyze very large data and detail trends, patterns, and associations, all involve algorithms, and are used in product recognition such as Netflix and Amazon and even vision and speech recognition.

Language is the "genetic endowment" of mankind but language alone does not explain the cognitive functions and abilities of the mind. Computers use a binary code but what computers lack is the ability to think abstractly and in many cases to make deductive associations. It is true the more data fed into a computer the better it can approximate something, but you can't teach a computer much about a phrase such as, "Physicist, Isaac Newton," even if we can make a search engine that returns sensible bits to users. Certainly, a computer can process information much quicker than you and I but a computer or robot does not have consciousness. We have already seen how much DNA resembles a computer, although I would better characterize it as how much a computer imitates DNA. The sequencing revolution with its high-throughput sequencing technique, aided of course by computers, has made reading DNA less expensive and allows scientists now to read the genome or genetic code for organisms much more quickly than in the past. One day, computers will personalize medicine, hinging on the ability to deal with masses of unanalyzed data, which is extremely difficult if not impossible for humankind to do. Of course, the key here is something like this requires design and exquisite design at that. Though deferring to naturalistic causes, Chomsky believed humans at "some point" emerged with new properties that other organisms don't have that has "arithmetical properties" which allowed language to develop, but he is well aware that there is absolutely no evolutionary cause for this.

For some, there is much concern that artificial intelligence will ultimately be the doom of mankind. Elon Musk has predicted that in five years something very serious will occur killing many people as a result of robots. Dennis Pamlin and Stuart Armstrong of Global Challenges Foundation and *TheGuardian.Com* believe a computer with human level intelligence "could not be easily controlled (either by the group's reality then or by some international regulatory regime) and would probably act to boost their own intelligence and acquire maximal resources for almost all initial AI motivations." Even Stephen Hawking thinks, "...the development of full artificial intelligence could spell the end of the human race." I don't know how realistic these worries are of artificial intelligence, but what I do know is neither machine nor man will destroy the earth; this has been left up to God (2 Pet. 3:12).

A bit different than artificial intelligence, artificial life has also been created in the lab. For example, the "Open Worm Team," part of a global team from California and the United Kingdom, produced a copy of the worm *C. Elegans* by implanting a digital "mind" into a Lego machine with 1000 cells recreated from the worm's brain with these neurons able to "fire" to make decisions. This artificial nematode mimics the action of a real worm like running into things and turning, and could be a precursor to a more complex animal machine but these machines cannot self-replicate which, of course, is a prerequisite for life.

One final thought. Mankind continues to attempt to create life that is biologic life yet thousands of experiments have failed to remotely produce any new or novel life form. From my perspective, if life is truly random and blind and has only come about by shear chance, it would seem likely an intelligent mind such as the human mind could figure out how to create life quite easily, especially with the sophisticated computers and brilliant scientific minds mankind has developed. The fact is, however, to date man has yet to produce anything that even approaches the most rudimentary form of biologic life.

Chapter 9

Birds, Moths and a Missing Link

Darwin's Finches

Among the most famous icons of evolution are the so-called "Darwin's finches" of the Galapagos Islands. Some thirteen species of finches were identified by Darwin, who theorized these finches had begun as a single species that then evolved into first several varieties and then, through natural selection, the distinct species. The species varied in size and color but what intrigued Darwin was primarily the differences in the size and shape of their beaks.

Surprisingly, the finches had little to do with Darwin's formulation of his theory, though this is seldom mentioned in textbooks. This is because they were only mentioned passingly in his *Beagle* journal and never in *On the Origin of Species*. Furthermore, these same species of finches are now reversing direction and emerging into fewer species!

So why are Darwin's finches given so much preeminence in spite of his sparse knowledge of their habits and geographical distributions? The Galapagos finches were not elevated into their iconic status until the rise of "Neo-Darwinism" in the 1930's. First called Darwin's finches by Percy

Lowe in 1936 and popularized in 1947 by David Lack, the Galapagos finches were given evolutionary significance, thus perpetuating the myth that Darwin's theory was based on this fact. As Frank Sulloway has noted, "Nothing could be further from the truth."

But is there evidence of

evolution concerning these birds? By now in our study you should be able to readily answer this. The answer is an emphatic no! Though natural selection could explain some variety in these finches, there is no change in kind. The finches are still birds and this is where the definition of species becomes so important. As man defines species, there may be thirteen species of finches on the Galapagos Islands but there remains only one kind. This is especially obvious when we observe the reversal of the shape and size in the beaks of these finches that occur when weather circumstances of the islands change. Yet, because they can interbreed, it is evident that many of them have not even completed the speciation process. After a long drought in the 1970's, followed by heavy rains in 1983 made possible by El Nino, the shape of their beaks changed. Joseph Weiner noted, "Selection had flipped. The birds took a giant step backwards, after their great step forward." This is also supported by hybridization. "At least half of the finches are known to hybridize," so said Joseph Weiner. This allows for "fusion" or interbreeding into the population, once again confirming the concept of kind.

Despite all this, the National Academy of Sciences still refers to Darwin's finches as "a particularly compelling example" of evolution. This is at least a distortion and absolutely offers no evidence for molecules-to-man evolution.

Peppered Moths

Is there a high school biology student who has not heard of the peppered moths and their alleged proof of evolution? Almost every textbook in Amer-

ica still alludes to the peppered moth as a classical demonstration of natural selection, the mechanism for evolution. The question we ask once again is: What are the facts concerning peppered moths and do they offer any evidence for evolution? Jonathan Wells in his book, *Icons of Evolution* has chronicled the history of the peppered moths' saga. It begins with an experiment carried out by a physician-biologist, Bernard Kettlewell, in a forest near Birmingham, England. The peppered moth, *Biston betularia,* has various shades of gray. Many years ago the "typical" peppered moth was predominantly a light gray. After a passage of time, however, the peppered moth became more "melanic" or in other words a darker shade of gray. In 1896, J.W. Tutt, "a biologist," theorized this change could be a result of the industrialization that had occurred in Birmingham. This was because the dark moth would be better camouflaged against the darker trees now made so by pollutants in the air.

As a result of this, Bernard Kettlewell released moths on the trunks of these polluted trees near Birmingham. Kettlewell's moths included the lighter or typical varieties as well as the darker melanic varieties. Once released, he set up traps to capture the moths. Of 447 melanics he recaptured 123 but of the 137 typical released he captured only 18, a 14% difference between the two. For this difference Kettlewell concluded, "Birds act as selective agents, as postulated by evolutionary theory." Kettlewell went on to reproduce his experiment in other parts of England with similar results thus causing him to proclaim "the most striking evolutionary change ever witnessed in any organism," this industrial melanism of peppered moths. As recently as 1975, P.M. Sheppard called Kettlewell's observation, "The most spectacular evolutionary change ever witnessed and recorded by man, with a possible exception of some examples of pesticide resistance."

Kettlewell's experiment had a problem though, as pointed out by Jonathan Wells. First, if these dark or melanic moths had such a great advantage over the typical ones, they should have completely replaced them. This of course never happened. There were also discrepancies with the data. For example, in rural Wales, away from pollution, the melanic form was higher than expected, for which there was no evolutionary explanation. These inconsistencies were found in other parts of England as well, causing biologists to evoke "nonvisual selective factors." Kettlewell and others tried to explain this by the differences in lichen on the trees, but this could not be correlated to other places such as America. Wells concludes that Kettlewell and other scientists made a serious error, and that is: they failed to look at the natural resting place for the moths, which turns out not to be the trunk of trees!

In 1980, Finnish zoologists showed peppered moths do not normally rest on the trunks of trees but instead beside them, more or less horizontal to the branches. Why is that significant? To prop up Darwinian Theory, the classic resting place of peppered moths is on the tree trunks. Moths rarely fly during the day; they usually remain where they are on the tree. If on the trunk, the melanic moth would have more of an advantage from a natural selection point of view but pictures of moths on tree trunks were staged.

They were placed there by either being pinned or glued and Kettlewell would later even admit to this. This does not necessarily mean that camouflage is not an advantage for the moth but it certainly calls into question the integrity of the biologists responsible for these pictures, especially when these pictures are still portrayed in biology textbooks. In the textbook we have already alluded to in our study, Miller and Levine's *Biology*, these fake photographs are still shown with Kettlewell's work called "a classic demonstration of natural selection in action." Though Miller and Levine may be forgiven if they were ignorant, other textbook writers knowingly use these fudged photos unapologetically, not wanting to make it too convoluted for young audiences!

The point here is just like the finches of the Galapagos, the peppered moths of England offer no support for molecules-to-man evolution, even if natural selection does favor a particular morphologic difference within kinds.

Archaeopteryx

Since Darwin's publication of *On the Origin of Species*, scientists have been searching for "the missing link." In actuality, there should be and must be tens of thousands of missing links for evolution to be true. Every kind should have evolved from another kind, hence many missing links that transform into the new kind.

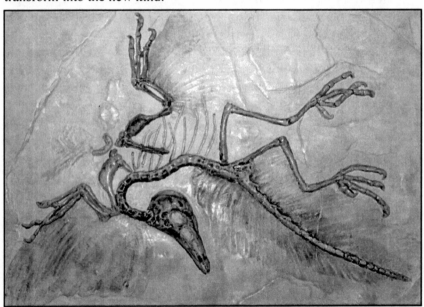

Archaeopteryx

Darwin knew of no missing links in the fossil records but as we have noted he attributed this to the imperfection of the fossil record. He was convinced scientists would eventually unearth these "transitional forms." Sure enough, in 1861, two years after Darwin's publication, a German scientist, Hermann von Meyer, found a fossil with wings and feathers but also teeth and a long, lizard-like tail. A more complete specimen of this creature was found in 1877. It was eventually transported to the Humboldt Museum in Berlin and has come to be named *Archaeopteryx*, a supposed link between birds and reptiles. It has been called "the most important natural history specimen in existence."

Although some have questioned the authenticity of the fossil remains of the *Archaeopteryx*, everyone now believes these specimens are genuine. But is *Archaeopteryx* "the almost perfect link between reptiles and birds" as Harvard Neo-Darwinist Ernst Mayr has proposed? Many paleontologists disagree. Larry Martin points out, "*Archaeopteryx* is not ancestral to any group of modern birds" because of too many structural differences between the two.

Controversy exists between evolutionists as to how flight evolved, some believing in a "tree down theory" and others in a "ground up theory." In other words, flight could have evolved by animals swinging from tree to tree or from animals running after their prey with flight becoming a distinct advantage. This would imply very different ancestors for *Archaeopteryx*. The prevailing theory now amongst scientists is that *Archaeopteryx* evolved from two-legged dinosaurs. Ironically, with this new view paleontologists were forced to conclude that the most likely candidates for precursors to *Archaeopteryx* lived tens of millions of years later! This then necessitated a rearrangement of the fossil remains with the obvious objection that an animal cannot be older than its ancestors. Jonathan Wells states that as a consequence *Archaeopteryx* must be removed as the first bird and become another feathered dinosaur.

The question then is: Where is the first bird? In the search for the supposed first bird, sometimes paleontologists have been the victims of fraud. For example, in 1999 *National Geographic* featured *Archaeoraptor*, a supposed flying feathered reptile. Later, through the efforts of Chinese paleontologist Xu Xing, *Archaeoraptor* was discovered to be a dinosaur tail glued to the body of a primitive bird. Another hoax was *Bambiraptor*, a chicken-sized dinosaur supposedly 70 million years younger than *Archaeopteryx* and called, "the most bird-like dinosaur yet discovered." Problem was nothing remotely resembling feathers was found with this fossil though these have been fabricated by ardent evolutionists.

To illustrate another feeble attempt by eager paleontologists, Wells points to a symposium where William Garstka and his team from the University of Alabama had found bird DNA in a fossil 65 million years old. After analyzing the DNA, they found "The first direct genetic evidence to indicate that birds represent the closest living relatives to dinosaurs." Once again, there was a problem. First, the DNA they found was from Triceratops, not a flying dinosaur. Second, and even more startling, this DNA was exactly the same as the modern turkey! Not just 99.9% but 100% turkey. How utterly foolish! As Wells aptly put it, if you are going to fake something, don't make it so pitifully obvious. For Wells, who attended the symposium, this was very enlightening. "The incident convinced me that some people are so eager to believe that birds evolved from dinosaurs that they are willing to accept almost any evidence that appears to support their views."

For evolutionists, the cladists who classify organisms dethroned *Archaeopteryx* as the missing link. Biblical creationists already knew this. No one doubts the unique features of *Archaeopteryx*. Whether you think it should be considered a flying dinosaur or a unique bird, there is no evidence to think of it as a missing link. Unfortunately, as you may have guessed, *Archaeopteryx* is still featured as a classic example of the missing link in textbooks and the taxpayer-funded Smithsonian Museum of Natural History, even though paleontologists now agree to the contrary. What a shame!

Eohippus

Evolutionists frequently speak of thousands of transitional forms but the truth is they only put forth a few examples. Transitional series have been proposed for shaled squid, oysters, and the horse, or Eohippus. The other two examples are clearly not transitional forms but rather examples once again of genetic variability. But is the horse series really an example of evolutions with distinct changes in kind?

When examining the Eohippus series or for that matter any series, three things should be considered: (1) change in morphology or body type; (2) change in size; and (3) where the fossil is found in the geologic strata. For the horse, the series shows increasing size, a change in the shape of the head, and most importantly a change in the hoof, going from four appendages to one.

In the beginning of the series as depicted in typical drawings, we see a small animal the size of a small or medium sized dog with four toes, progressing to a large sized animal beginning to look more like a horse but

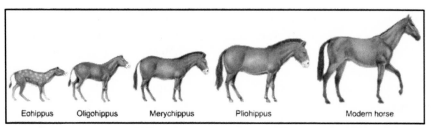

Eohippus　　Oligohippus　　Merychippus　　Pliohippus　　Modern horse

with three toes, eventually culminating in the modern horse with one hoof. Of course, those are supposed to represent fossils that have been found, but interestingly no fossil has been discovered of a two toed animal, the gap being filled in by evolutionists. The series we see in textbooks is apocryphal, it does not exist except in textbooks!

When looking at the horse series, we see the first animal depicted looks very much like an animal that exists today, the Hyrax. This animal has four toes and can get as large as a medium sized dog.

One must also keep in mind genetic variation. Modern horses come in many sizes and with many morphological differences. Miniature horses can be bred to be as small as a dog with heads that can look very different. It is clear that size and morphology alone cannot be used to distinguish a transitional series. Even the number of hoofs can be a result of genetic variability. When looking at today's horse hoof we see one large hoof with two other structures (heel buds) that may have served as hoofs before. Of course, it could be and probably is that the four and three toed animals could simply represent different kinds with evolutionists' wishful thinking explaining the supposed transitional series.

But there is one other problem with the Eohippus series, and that has to do with the strata where the fossils have been found. Recall the concept of the geologic column, where ancient animals should be found in lower strata with more modern ones found in the higher or more recent strata. In the case of Eohippus, the one hoofed animal is found in the same strata as the three hoofed and in South America the one hoofed is found in lower strata than the three hoofed. This obviously makes no evolutionary sense.

Of course, there should be transitional forms for all modern organisms but there simply are not. Some may ask, how about the woolly mammoth and the elephant, or the saber tooth tiger and the modern tiger? These and others like them are examples of genetic variation, with extinction of the former and survival of the latter by natural selection. Once again, this is adaptation but adaptation does not result in new kinds! The modern horse does not look like it did 5,000 years ago, neither does the elephant or the tiger or for that matter, even humans. Natural selection is the reason for this and creationists are fully cognizant of it. This term "natural selection" is not necessarily a bad phrase; it is the mechanism for adaptation through genetic variability.

Chapter 10

The Human Fossil Record

The ultimate icon of evolution is the typical picture you find in either a *National Geographic* magazine or any high school textbook of biology showing the evolution of man beginning as a lower primate or chimpanzee, eventually through a series of changes becoming bipedal, turning into what ultimately looks like a caveman and finally the picture of a modern man. These pictures are unapologetically offered as evidence for man's evolution supposedly derived from the fossil record. In fact, this parade representing the ascent of man is purely fictional. It has no basis in reality. Unfortunately, the pictures do have the ability to leave an indelible image especially when they are presented as fact. In this chapter, we are going to look at the fossils that have been found by paleontologists that supposedly give evidence for the evolution of man through lower animals such as chimps. What one must understand, however, is that the fossil records are there primarily to serve the evolutionist. In other words, when fossils are discovered they are interpreted from an evolutionary presupposition and are placed in order according to evolutionary theory. Creationists, such as myself, reject that mankind has evolved from lower animals, not only from a biblical point of view but also from an intellectual point of view. We will find that the fossil record at this point in time simply does not support evolution and there are no compelling transitional states or missing links.

There are a few things that we need to keep in mind, however, when looking at the fossil record and the interpretation of that fossil record by paleontologists and paleoanthropologists. First, these fossils and fossil fragments are treated as almost sacred, as if they are some sort of religious relics by the curators of the museums and universities that house them. In fact, you will virtually never see the actual fossil remains at a museum; they are always reproductions. Even university graduate students wishing to study the actual human fossils will rarely, if ever, be allowed to do so. Author

Marvin L. Lubenow, in his book, *Bones of Contention*, states, "No prisoner on death row is under greater scrutiny than those ancient relics called human fossils." The second point is some of these fossils have actually been lost and are no longer available except in their plaster casts. Thirdly, one must also keep in mind that the fossils can be arranged in an alternative manner than how the evolutionists arrange them, which we will discuss further in this chapter. Consider also that the drawings that are reconstructed from these fossil remains always go beyond what could really be intelligently derived by simply looking at skeletal remains.

Jonathan Wells in his book, *Icons of Evolution*, has given a good example of how ludicrous it would be to trust an artist's rendition made on fossil remains. He states, "Just recently, *National Geographic* magazine commissioned four artists to reconstruct a female figure from casts of seven fossil bones thought to be from the same species as skull 1470 (1470 represents fossil remains discovered in Kenya in 1972 and studied by Alan Walker, Michael H. Day, and Richard Leakey). One artist drew a creature whose forehead is missing and whose jaws look vaguely like those of a big dinosaur. Another artist drew a rather good-looking modern African-American woman with unusually long arms. A third drew a somewhat scrawny female with arms like a gorilla and a face like a Hollywood werewolf. And a fourth drew a figure covered with body hair and climbing a tree, with beady eyes that glare out from under a heavy, gorilla-like brow."

Yet another thing to keep in mind is the obvious, that being, paleontologists assume the evolution of man. Similarity, therefore, equals evolution, which goes back to the homology argument that we have already reviewed.

The true fact is, as stated by Marvin Lubenow, "A series of objects created by humans (or God) can be arranged in such a way as to make them look as if they had evolved when in fact they were created independently by an intelligent being." The problem we see with paleontologists in constructing the evolutionary ascent of man, especially pictorially, is that when studying the myriad of fossil remains that have been discovered, fossils are selectively excluded if they do not fit well into an evolutionary scheme. Also, some fossils are arbitrarily downgraded to make them appear to be evolutionary ancestors when they are in fact true humans, and finally some non-human fossils are upgraded to make them appear to be human ancestors.

Again, from Marvin Lubenow, "The myth in the minds of the public is that human fossil material is readily available and is thoroughly studied by all who teach and write on the subject. The truth is that paleoanthropology is in the awkward position of being a science that is several steps removed from the very evidence upon which it claims to base its findings." In other words, to think that paleontology and paleoanthropology is a true hard science would be a mistake. Carl Sagan once said, "Not all scientific statements have equal weight." Andrew Hill of Yale University observes, "Every new fossil hominid specimen is the most important ever found and solves all known phylogenic problems; every new hominid specimen is completely different from all previous ones, no matter how similar; every new hominid specimen is a new species and probably a new genus, and therefore deserves a new name." This type of phenomenon is simply not seen in the true natural sciences. Mr. Hill goes on to note, "Compared to other sciences, the mythic element is greatest in paleoanthropology." Milford Wolpoff states, "When the only people who can comment are the discoverers or friends of the discoverers, there is no sense of independent observer. We are not practicing science. We are practicing opera."

In 1991, Yale graduate Misia Landau, in her book *Narratives of Human Evolution*, maintained that many "classic texts in paleoanthropology" were "determined as much as by traditional narrative frameworks as by material evidence." The typical stories, showing human ancestors moving from trees to ground and then developing upright posture, acquiring intelligent language and developing technology in society, "far exceed what can be inferred from the study of fossils alone and in fact places a heavy burden of interpretation on the fossil record – a burden which is relieved by placing fossils into pre-existing narrative structures," says Landau. Furthermore, Ms. Landau has recognized the similarities between the stories of human evolution and the stories of folktales and hero myths of literature. In both instances, there is an initial condition of the hero, in evolution's case the helpless, defenseless primate living in a tree. Then comes a change in circumstances where, again, in the evolutionary story finds the hero developing a more upright or bipedal structure and a larger brain. Then the hero becomes tested, in our scenario by predators or harsh climate. As a result there is a transformation and the "hero" acquires increasing intelligence and tools. The hero becomes tested yet again, eventually evolving into full-fledged human beings. Finally, the hero will succumb. In almost all evolutionary tales, human evolution ends with man's ultimate destruction primarily derived from his own technology.

Misconceptions

There are a few misconceptions regarding fossil remains that should be pointed out before we proceed. The biggest misconception is that there are relatively few fossils that have been studied in regard to the evolution of man. In fact, there are numerous fossils that have been unearthed. So when workers in the field speak of the scarcity of human fossils, what they are actually saying is that although there is an abundance of such hominid fossils, most of them are either too modern to help them or they do not fit well into the evolutionary scheme. And since these paleontologists feel that humans must have evolved, it is perplexing and difficult to understand why they haven't found enough fossils that would "clearly demonstrate that fact." Another misconception to keep in mind is that science is honestly self-correcting. Let's take the case of the so-called Piltdown Man. Charles Dawson in 1908 and in 1915 supposedly found evidence for the missing link and it came to be called the Piltdown Man, Piltdown referring to a tiny hamlet in East Sussex, England where the skull, mandible and tooth, and ancient tools, were claimed to have been found. In 1953, however, some 38 to 45 years later, it was finally discovered that Piltdown Man was a fraud. Although the skull found may have been real, the mandible and tooth were from an orangutan and further the tools found there had been "placed there." You might say science eventually got it right but during the interim Piltdown Man was lauded as the true missing link by paleontologists of that time.

Falsification

Before moving on, I would like to address the concept of falsification. Any worthy scientific theory must be capable of being falsified and since human evolution is a historic process this falsification must come from the fossil record. If an alleged human ancestor's fossil records were found to exist and be contemporaneous with human or *Homo sapiens* records, since one allegedly evolved into the other, that evidence would falsify the theory. That is precisely what we shall discover in our study. For evolutionists to ignore it indicates evolution is more philosophical than scientific. In 1974, Donald Johanson in Ethiopia discovered a supposed 3 million year old female fossil he described as "our oldest ancestor" and gave the name to his finding Lucy. Supposedly, Lucy was the product of 5 million mutation events and 3 million years and even with this time required "punctuated evolution" in order to develop. Similar skeletal findings became known as *Australopithecus afarensis*. As mentioned, Johnson dated his specimens at 3 million years. T.C. Partridge, on the other hand, discovered other *Australopithecus africanus* fossils he dated at 870,000 years, far too recent for the comfort of paleontology. These were considered "nonconformist fossils." The interesting thing about Partridge's find was that modern hominids were already on the scene, according to paleontologists, as early as 750,000 years ago. So how do evolutionists deal with this conflict? First, consider this primary fact, for evolutionists, in this case the fossils described by Mr. Partridge were upgraded to *Homo habilis* or a more advanced creature than *Australopithecus africanus*. In other words, paleontologists waved their magic wand. Marvin Lubenow has pointed out it becomes impossible to **falsify** a theory when the scientists can change their beliefs in midstream. It is like trying to nail jelly to a wall. Bad data for the evolutionists is data that doesn't fit their already certain theory. The oldest human fossil supposedly ever found was the KNM-KP-271 fossil from Lake Rudolf in Kenya near Kanapoi, supposedly 3.5 million years old. The fossil is shaped exactly like *Homo sapiens* but because of its supposed age was called *Australopithecus africanus*. The fact is *Australopithecus afarensis* including Lucy, and *Australopithecus africanus* were not human at all, but represent extinct primates.

Neanderthal Man

In 1856, in the Valley of Neanderthal, named after Joachim Neander from the Middle Ages in Germany, was found the fossil remains of what would become called "the Neanderthal Man." Since 1856, many other skeletal remains, in the neighborhood of 200, have been found of Neanderthal Man. At the time Neanderthal Man was considered to be ancestral to modern humans. When, at the time, Dr. Rudolf Virchow of the University of Berlin, the Father of Pathology, examined the remains he was convinced they were *Homo sapiens* with rickets! Even Thomas Huxley, Darwin's Bulldog, agreed that Neanderthal Man was indistinguishable from *Homo sapiens*. Neanderthal Man supposedly goes back 800,000 years ago. The problem evolutionists have with Neanderthal is they don't know where he came from and they don't know where he went as apparently he disappeared some 34,000 years ago. For reasons we will discuss later, the present popular view by paleontologists is that the Neanderthals represent an isolated side branch on the human family tree. Neanderthals' distinct morphology includes large cranial

Neanderthal Man

capacity **averaging that of modern man**, a skull shape low, broad and elongated, the rear of the skull somewhat pointed like a bun; large heavy eyebrows, low forehead, large long faces with the center of the face jutting forward, weak rounded chin, and post-cranial skeletal bones that are very thick, as well as short stature. These morphological changes can all be explained by non-evolutionary processes such as geographic isolation and genetic recombination. Health factors including vitamin D deficiency as well as infectious disease could also explain the morphology of Neanderthal. Evidence now exists that show Neanderthals had tools, jewelry or ornaments, and hearths; lived in structured or walled areas; had boats that were used for fishing; and even had musical instruments. Neanderthal Man then was contemporaneous with modern humans and fully human culturally as well. Modern DNA evidence indicates that Neanderthal and Homo Sapiens interbred to produce offspring, thus they were the same species.

Java Man

Java Man

Java Man has been called the "true ape-man." Originally referred to as *Pithecanthropus erectus* (erect ape-man), now referred to as *Homo erectus*, he was first discovered by Dutch anatomist Eugene Dubois, in the Dutch East Indies along the bank of the Solo River in Java, in Indonesia. His first find was that of a skull cap which he believed had a combination of human and ape features, and a year later about 50 feet away he found a femur that was very human in appearance. Dubois, assuming these belonged together, thus constructed his Java Man which is now virtually synonymous in the popular mind with human evolution. For reasons we will discuss shortly, Marvin Lubenow believes Java Man is not our evolutionary ancestor but a true member of the human family and a smaller version of Neanderthal.

Part of the problem regarding Java Man is the dating given by Dubois. Dubois, who was an anatomist, not a geologist, was not qualified to date the strata in which the skull was found and probably significantly overestimated the age. Also, there was a problem in the collection of the data by Dubois. Dubois himself did not uncover any of the important fossils ascribed to him; they were found by those working with him and brought to him for his examination. Further studies by paleontologists of the femur found by Dubois conclude it is indistinguishable from that of a human femur, if, in fact, the skull cap and the femur actually do belong together. Finally, the skull cap, by many paleontologists, is considered similar to Neanderthal with the capacity of 1,000 CC. All things considered it certainly appears Java Man was human and part of the *Homo sapiens* category.

Wadjak Man (Wajak)

Eugene Dubois was not through with his excavations. Travelling back to the Dutch Indies he found the skulls of another creature that was very similar to *Pithecanthropus*. These skulls had a capacity of 1550 cc and 1650 cc and for all intents and purposes appeared to be truly human. The reason you may not have ever read about Wadjak Man is that Dubois' work was never published. He hid it from the public because of the significant similarity that it had with *Pithecanthropus* and he needed to "save *Pithecanthropus*" thus saving his own reputation. No paleontologist now doubts that Wadjak Man, which is dated approximately 10,000 years ago, is anything other than human. An interesting finding in the excavation of Wadjak Man was the finding of a fossil remain of *"Tapirus indicus,"* a tapir that was supposedly extinct millions of years before.

Selenka-Trinil Expedition

In the early 1900's, Professor Emil Selenka was preparing to visit the site of Dubois' earlier expedition where Java Man was found. Unfortunately Professor Selenka died before this could be carried out, but his wife, Frau M. Lenore Selenka, organized another expedition to Java to explore where the original *Pithecanthropus* had been found. The expedition became known as the Selenka-Trinil Expedition, Trinil referring to the area in Java where the expedition was carried out. What came forth was a 342-page document describing the findings at Trinil. Although extensive diggings and mining were carried out, no *Pithecanthropus* was found. There were many scientists involved, which included specialists in geology who estimated the beds to be only 500 years old. Interestingly, what was found at the Trinil site was a human tooth. Unfortunately, the well-documented Selenka-Trinil expedition has been largely ignored by the evolutionary and paleontology community in what Lubenow refers to as "an amazing conspiracy of silence." The reasons are obvious why the findings have been ignored, in that they contradict Eugene Dubois and would also put *Homo erectus* or *Pithecanthropus* in a much later evolutionary time.

So is *Homo erectus'* morphology distinct enough to warrant it being considered a separate species from *Homo sapiens*, and are there non-evolutionary ideas to explain the morphology such as we saw in the Neanderthal Man? The answer is no, there is not enough morphologic difference to warrant putting *Homo erectus* in a separate category from *Homo sapiens*, and yes, there are geographic reasons for the difference in morphology just as we discussed with Neanderthal Man. *Homo erectus* is in fact becoming the evolutionist's worst nightmare. The problem is some of the fossil remains are dated very old and some are dated very young, ranging from 1.5 million years to 27,000 years old. Paleontologists have trouble with these dates. The obvious concern they have is that for *Homo erectus* to coexist with humans, which these dates would allow, would not constitute how evolution is supposed to work. Furthermore, for *Homo erectus* or any other species to exist for 2 million years without significant evolutionary change would also violate evolution. Using the dating system accepted by evolutionists, there are now at least 72 fossil records of *Homo erectus* that are less than 30,000 years old, with the youngest being 6,000 years old. It is generally agreed by paleontologists that *Homo sapiens* have been around at least 10,000 years. This is significant overlap and clearly shows that *Homo erectus* coexisted with man because he was man, albeit in shorter form, which genetic variability allows.

The same things can be said of *Sinanthropus pekinensis* (Peking Man, now *Homo erectus*). The original cast of Peking Man was lost and what we have now is simply a fine plaster cast. The skull size is compatible with other *Homo erectus* remains. William S. Laughlin of the University of Connecticut has studied the Eskimos and the Aleuts extensively and has noticed similar morphologic features between these people and *Homo erectus*, or Peking man. He concludes, "We find that significant differences have developed, over short time span, between closely related and contiguous peoples, as in Alaska and Greenland, and we consider the vast differences that exist between remote groups such as Eskimos and bushmen, who are known to belong within a single species of *Homo sapiens*. It seems justifiable to conclude that *Sinanthropus pekinensis* (Peking Man) belong within the same diverse species."

The differences between *Homo erectus, Homo sapiens,* and Neanderthal can all be explained by genetic variability. Looking at modern day human beings this variation is obvious. Cranial capacities range from 700 cc all the way to 2200 cc. The cranial capacity in *Homo erectus* ranges from 780 cc to 1225 cc and the cranial capacity of Neanderthal remains begin at 1200 cc and go to about 1650 cc. Other than the differences in cranial capacity between them and modern man there are virtually no other morphological differences. What can be concluded is that *Homo erectus* and Neanderthals constitute what we call *Homo sapiens* or humans and there is no reason to categorize them otherwise. Clearly, *Australopithecus africanus* and *afarensis* as well as *Homo habilis* were not human, and would belong to the ancient ape or chimpanzee category.

Archaic *Homo Sapiens* and the Laetoli Footprints

There are a group of fossils from Europe, Africa and Asia that have been called unofficially "archaic *Homo sapiens*." They do not fit into either the Neanderthals or *Homo erectus* categories because: (1) They have different skull morphology from the classic Neanderthal; (2) Many of these are dated much earlier than the classic Neanderthal; and (3) The cranial capacity is too large for them to be classified as *Homo erectus*. The dates vary concerning these archaic *Homo sapiens* from as recent as 5,000 years ago to as late as 700,000 years ago. It is clear that all these groups are fully human and part of the human family created by God. In fact, creationists have felt the dating of the archaic *Homo sapiens* disproves the concept of evolution.

Ian Tattersall believes these fossils presented a problem for the evolutionists. He states, "The hominid fossil record of the past 300 to 400 thousand years offers a remarkable degree of morphologic variety. Yet (late persisting *Homo erectus* aside) conventional wisdom assigns all these fossils to *Homo sapiens* albeit of archaic varieties."

In 1992, Juan Luis Arsuaga found undisturbed fossil deposits in a cave at Sima de los Huesos in Spain. Originally he found three well-preserved fossil skulls but since that time 33 have been discovered and dated by evolutionists at about 400,000 years. These remains show a wide degree of variations within a contemporaneous population. As Marvin Lubenow observes, this sort of "muddles in the middle" all the other European fossils that appeared to be so different. He notes the physical variation found in this one assemblage of fossils encompasses "all the other European archaic *Homo sapiens* fossils." They all have similarities to *Homo erectus*, Neanderthal and *Homo sapiens*. This has rather profound implications for a creationist. This extreme variation in population is exactly what we would expect. Again, from Mr. Lubenow, "The distinction made by evolutionists between *Homo erectus*, early *Homo sapiens*, Neanderthals and anatomically modern *Homo sapiens* now fades into insignificance." It is a remarkable affirmation of the biblical statement from Acts 17:26, "From one man he made every nation of men, that they should inhabit the earth."

It is obvious that all the variation in the fossils at Sima de los Huesos can be attributed to non-evolutionary explanations. It points to the absurdity of attempting to determine species distinctions in human fossils and contrary to what evolutionists may believe, non-evolutionary morphologic changes can take place very quickly due to genetic isolation or the ending of genetic isolation and resulting genetic recombination. What is clear is whether it involves Neanderthals or Australian aborigines or North American natives, it is extremely difficult to determine genetic relationship after thousands of years have passed. The only legitimate species criterion is a fertility test or a DNA equivalency!

The Laetoli Footprints

In 1978, Mary Leakey discovered a series of what appeared to be human footprints at Site G, Laetoli, about 30 mile south of Olduvai Gorge in Northern Tanzania. The footprints were dated an alleged 3.6 million years and Ms. Leakey described them as "remarkably similar to those of modern man." The trail of footprints contained a total of 69 prints extending approximately 30 yards in length. As you might guess, these footprints have been the subject of a large body of literature. They have been described as Lucy-type hominid or *Australopithecus afarensis* despite the striking appearance of modern humans. Of course, this is completely unprovable and even more curious, the footprints are the size of those made by

The Laetoli Footprints

modern humans; but Lucy was only three feet tall. Russell H. Tuttle of the University of Chicago has studied the Laeotoli footprints extensively. In his studies he realized there were a few studies done on habitually unshod peoples. As a result he observed the Machiguenga Indians of the mountains of Peru. These people are habitually barefoot. In examining their footprints he concluded, "In some, the 3.5 million year old footprints at Laetoli Site G resembled those of habitually unshod modern humans. None of their features suggest that the Laetoli hominids were less capable bipeds than we are." So why not ascribe those footprints to humans? Mr. Tuttle answers this question himself, "If the G footprints were not known to be so old, we would really conclude they were made by a member of our genus Homo." In essence, then, to ascribe those footprints to humans does not fit the evolutionary timescale. As Mr. Lubenow points out, "Interpreting the Laetoli footprints is not a question of scholarship; it is a question of logic and the basic rules of evidence." Evolutionists will never call extremely old fossils by their proper names because to do so would invalidate their theory.

The facts are fossils that are indistinguishable from modern humans can be traced all the way back to 4.5 million years according to the evolutionary timescale. That would suggest there were true humans on the scene before Australopithecines appear in the fossil record. So even when we accept the evolutionists' dates for fossils (which I don't) the results do not support human evolution.

One solution to this problem is to now claim that fossils are not really all that important. Mark Ridley of Oxford University states, "No real evolutionist, whether gradualist or punctuationist, uses the fossil record as evidence in favor of the theory of evolution as opposed to special creation. The evidence for evolution simply does not depend on the fossil record." To be sure, Darwin said the number of intermediate forms fossilized in the records must be numerous and if they are not found their absence posed a serious dilemma for his theory. After 150 years this continues to be the case. Marvin Lubenow has stated, referring to evolution, "It must be the only theory ever put forth in the history of science that claims to be scientific but then explains why evidence for it cannot be found." The human fossil evidence is in complete accord with Scripture.

Chapter 11

The Cambrian Explosion

Much has been written about the so-called Cambrian Explosion by the scientific community. For creationists or the Intelligent Design scientists, primarily using the Cambrian layer of the geologic column shows how life suddenly occurred with complexity and with no evidence of gradualism, which of course is an evolutionary prerequisite. Naturalists and evolutionists deny there was any type of "explosion" and ostracize creationists for overstating what has really been found in the Cambrian geologic column. So are the facts concerning the Cambrian fossils supportive of a creationist explanation?

First, let's consider the Cambrian layer. We have already discussed the geologic column and as you recall there is nowhere on earth where we find an absolute complete column. Nonetheless, scientists use the geologic column to date fossilized animals found in them. Evolution predicts the older layers will have more primitive animals and that these animals become more complex as the geologic layers approach our current time.

Twenty-eight years before Darwin's book *On the Origin of Species* was published, Adam Sedgwick, an English biologist, had studied some curious fossils that had been discovered in Northwestern Wales that would later be termed Cambrian fossils (Cambrian being a Latinized English term for the country Wales). In fact, Sedgwick had actually taken Charles Darwin with him on his exploration of the upper Swansea Valley in Wales. Cambrian then would be a designation that would eventually replace Silurian for the earliest strata of animal fossils. Years later, very similar fossils were discovered in the Burgess Shale Formation in British Columbia by Charles Walcott in 1917. It would later be studied by several more biologists and paleontologists, including Harry Whittington of Cambridge in the 1960's. The Burgess Shale formation is 7,500 feet from a previous sea floor, having been eventually deposited in the Rockies secondary to the shifting of plate tectonics.

The Cambrian layer is one of the earliest geologic layers where life is found, although a few fossils of life have been discovered in the so-called Pre-Cambrian layer. The geologic column before the Pre-Cambrian layer contains no life at all. The Pre-Cambrian strata identify only three animal phyla and per evolutionary dating would have lived over 630 million years ago. The Cambrian layer, on the other hand, has cumulatively twenty-three animal phyla that are estimated to have lived 530 million years ago.

There is a puzzling pattern for paleontologists as it concerns the Cambrian period. First, there is a sudden appearance of animal forms that are complex, yet have no transitional intermediate ancestry. Second, the Cambrian animals are novel and have a wide variety of body plans. Thirdly, there are radical differences in the form of fossil records arising before more minor variations are seen. This clearly does not support Neo-Darwinian evolution.

You might be inclined to think that maybe the Pre-Cambrian animals served as early forms for the Cambrian organisms. The problem is: these fossilized organisms (of which there are very few) bear no clear relationship to any organisms that appear in the Cambrian Explosion, or for that matter any period thereafter. The most noted Ediacaran (Pre-Cambrian) organisms do not have obvious heads, mouths, sense organisms, guts or eyes, causing some paleontologists to question whether they should even be considered animals. So Pre-Cambrian animals offer no evidence of a fusing or transition into Cambrian animals and they, in and of themselves, are also a puzzling leap in biologic complexity with no transitional forms. Kevin Peterson of Dartmouth notes these Pre-Cambrian or Ediacaran fauna represent, "A quantum leap in ecological complexity as compared with the boring millions of years that characterized Earth before the Ediacaran."

Even so, the Cambrian period animals offer significantly more complexity when compared to the Pre-Cambrian period. It should be noted that there is no paucity of specimens to evaluate, as the Burgess Shale alone contains tens of thousands of fossils which can be found there today if one is inclined to make the long trek to that location. In fact, Charles Walcott himself had collected over 65,000 specimens during his exploration of the Burgess Shale. So what is the importance of the Cambrian fossils? As we have seen, Neo-Darwinian evolution requires random genetic changes from mutations to produce new variations for natural selection to work. The Ediacaran period supposedly lasted 94 million years, with the Cambrian lasting 55.6 million years. Though I don't agree with these dates, they still represent an exceedingly brief time for Neo-Darwinian mechanisms to work.

Both Pre-Cambrian, but especially the Cambrian, animals represent complex forms that look nothing like each other. We have already noted the Cambrian animals represent 23 phyla. Stephen Meyer in his book, *Darwin's Doubt*, points out, "The more an animal form manifests the characteristics of one phylum or group within the phylum, the less plausible it becomes as an ancestor of all the other animal phyla, and this is the dilemma evolutionists face in a nutshell." Again, from Meyer, "Cambrian phyla leave no evidence, by themselves, of the gradual emergence of the complex anatomical novelties that define the Cambrian animals."

Cambrian Animal Fossils

So how complex are the Cambrian animals? By far the most readily found and hence the most popular Cambrian animal is the trilobite. I am sure you have seen pictures of these if not the actual fossils themselves. They look like a miniature horseshoe crab and though they may not seem very complex to us, they actually are and seem to have arisen in the fossil records suddenly with no ancestral precursor. This animal had a head, complex eyes, gills, pleura, tail, and multiple legs, and represents organization and complexity of significance. Darwin would have realized that building a trilobite from a single cell organism using natural selection in a step-by-step manner would require many transitional forms encompassing a vast amount of time, for which there is simply no evidence. Other Cambrian animals are similarly complex and include the shrimp-like arthropod, *Waptia*; jellyfish-like *Eldonia*; worm-like *Phoronida*; and the

Cambrian fish *Myllokunmingia feng jioa*. Just like the trilobite, there is no evolutionary explanation for these animals.

So do evolutionists admit there is a problem? Actually they do, but they sure don't want creationists knowing they do! They will acknowledge the problems with their colleagues, but will minimize it with the public lest they seem to aid and abet the creationists. Some have tried to explain the lack of precursors and even Walcott himself did so. The most popular explanation is referred to as the artifact hypothesis, which Walcott espoused. Basically, the artifact hypothesis blames the lack of precursor Pre-Cambrian fauna on the concept of transgression and regression of ancient seas. Because of this transgression and regression Pre-Cambrian precursors would be inaccessible to discovery. This, in essence, is a negative argument attempting to explain away the absence of evidence, and Walcott himself realized this. He states, "I fully realize that the conclusions above outlined are based primarily on the absence of marine fauna in the Algonkian Pre-Cambrian (rocks) but until such is discovered I know of no more probable explanation of the abrupt appearance of the Cambrian fauna than that I have presented." In a version similar to Walcott's, some geologists suggest all Pre-Cambrian sedimentary rocks have been destroyed by heat and pressure in a process called "universal metamorphism." Others suggest that major bursts of evolutionary innova-tion occurred only during periods when sedimentary deposition had ceased; therefore resulting in a lack of fossil preservation. All these arguments have a common thread, as evolutionist Stephen Gould noted; they are "forced and *ad hoc* born of frustration, rather than the pleasure of discovery."

Finally, some have argued that soft bodied organisms that predated the Cambrian animals could not be fossilized due to their "soft-body biota." This argument was essentially laid to rest by Xian-Huang Hou, a Chinese paleon-tologist, who in 1984 studying fossilized samples in Southern China found many soft-bodied organisms. Hou notes, "The remains of hard tissues, such as shells of brachiopods or the carapaces of trilobites, are well represented in the Chengjiang fauna but less robust tissues, which are usually lost through decomposition, are also beautifully preserved." Since that time, even a greater variety of soft-bodied animals have been discovered in other deposits such as the Maotianshan Shales. "These discoveries demonstrate beyond any reason-able doubt that sedimentary rocks can preserve soft-bodied fossils of great antiquity in exquisite detail, thereby challenging the idea that the absence of Pre-Cambrian ancestors is a consequence of the fossil record's inability to preserve soft-bodied animals from that period." says Stephen Meyer.

For me, the issue is really much more problematic than just the Cambrian explosion. Even if you accept the concept of the geologic column, every period demonstrates the same phenomenon; that is the abrupt appearance of complex organisms without transitional forms, and this is paramount. As we have seen before, Darwin states, "If numerous species, belonging to the same genera or families, have really started into life all at once, the fact would be fatal to the theory of descent with slow modification through natural selection." Neo-Darwinism should be declared clinically dead!

Chapter 12

A Word about Entropy

In the past, entropy, or the second law of thermodynamics, has been used by creationists to disprove evolution. Since evolution is a violation of this law, therefore divine intervention must have occurred. On the surface, this would certainly appear to be true. Entropy can be defined as the amount of disorderedness (i.e. everything is moving from a state of order to a state of disorder, randomness, and confusion). The second law of thermodynamics states that in a closed system, entropy always increases, therefore, life would seem a contradiction. Evolutionists, as you might expect, deny that life violates the second law of thermodynamics. The question then is which one is right, and for that matter whether the second law of thermodynamics is, in fact, a non-variable law found throughout the universe. In order to establish this we must first examine the concept of entropy. We will then discuss how naturalists explain the way life does not violate this law, after which we will discuss how creationists say it does.

The second law of thermodynamics concerns itself with time, energy and disorderedness. The second law seems to be immutable and has been referred to as a metalaw on which different kinds of laws of physics can possibly work. The second law of thermodynamics was first formulated in 1824 by a French engineer named Sadi Carnot. Carnot was trying to develop the most efficient steam engine he could. He questioned whether it was possible to have a perfectly reversible engine, meaning entropy would stay constant, and concluded this was impossible. Forty years later, a German physicist, Rudolf Clausius, expanded on Carnot's original idea, realizing that what Carnot was describing was a law of nature.

Sadi Carnot

Clausius discovered that heat always flows in one direction. To clarify: when two objects of differing temperatures come in contact with each other, heat gradients tend to even out to reach a sameness or equilibrium between the two. Clausius invented the word *entropy* to explain exactly what was happening. Entropy for Clausius was the change of heat divided by the temperature, which always resulted in a positive number. Entropy is going from more order to disorder, and when this is accomplished, work is done. Something with low entropy, we would call fuel. Another way to contemplate the idea of entropy is that entropy increases because the arrow of time is always going forward.

In the 1870's, Ludwig Boltzmann defined entropy in a way that has become the accepted definition. In his definition, heat is actually thermal energy, the random motion of atoms. He realized the arrangements of atoms are macroscopically indistinguishable, and his major insight was that entropy is simply a way of counting the number of arrangements of atoms in a certain system. For Boltzmann, entropy was the number of possible microstates at which an entity can be. The reason entropy increases, according to Boltzmann, was simply that there are more ways to be high entropy than to be low entropy. Boltzmann mathematically formulized entropy where entropy equals Boltzmann's constant times the logarithm of the number of ways we can arrange things.

I know this is a bit technical, but it is important to grasp the concept of entropy before we proceed. Think of it this way. If you had a plastic candy Easter egg that was full of gas, therefore full of atoms, and you opened the egg, you would expect those atoms to escape and fill the room that you are in. This is going from low entropy to high entropy and this is how our universe seems to work.

For physicists interested in the cosmology of the universe, the universe began at a very, very low entropy state that they refer to as the Big Bang. Since the Big Bang, entropy has been increasing in the universe and it will until the universe reaches "equilibrium," which is also referred to as heat death or thermal death of the universe.

The obvious big questions are: Why was entropy low at the beginning of the universe and why was there a Big Bang to begin with? Physicists have been grappling with these two questions for many years and have postulated answers, but the bottom line is they just really don't know.

The question physicists must answer if we live in a truly naturalistic universe, is why life has occurred, and why has life changed from less complex to the more complex? It is clearly not a requirement of evolution that life must become more complex. No physical law demands this.

Their answer centers upon the concept that entropy decreases in small systems (life) and it does so because entropy is increasing on a much greater, wider scale! In other words, they admit life violates the second law of thermodynamics even though no experiment scientifically has ever shown any other system to do this. Physicists agree that the universe itself does not require entropy to increase; it is just an observable fact.

To muddle the picture a bit, physicists would ask us not to confuse complexity or simplicity with low entropy or high entropy. In the 1960's, a Soviet mathematician, Andrey Kolmogorov, defined complexity as the length of the shortest description that can capture everything relevant about a thing. Obviously, more complex entities require a longer description. Physicist Sean Carroll of the California Institute of Technology explains that complexity can arrive when entropy is in between low and high states. For example, if you pour cream into a cup of coffee, the cream will eventually disperse within the coffee turning the coffee a nice brown color. This is going from a low entropy state to a higher entropy state. But if you look closely, before the coffee turns brown there will be an intermediate phase where there will be variation of colors within the coffee. This would represent for physicists a more "complex" state even though entropy is increasing. When equilibrium is reached and the coffee is one solid color it is less complex but has a higher entropy. According to Carroll, complexity depends on entropy. In fact it relies on it. "The simple fact that entropy is increasing is what makes life possible," states Sean Carroll in his book *Mysteries of Modern Physics: Time*.

So what is life? It seems rather obvious to us. Previously we have shown that life is made up of DNA or RNA with the unique feature of life being self-replication. This is, however, not a satisfactory definition for most physicists. Erwin Schrödinger (a pioneer of Quantum Mechanics) defines life in this way, "It is something that goes on doing something longer than you would expect it to do." To demonstrate this, let's refer back to the cream in the cup of coffee example. As we stated, cream in a cup of coffee disperses and changes the color and the molecules interact and we have one solid color with higher entropy. Now, let's place a goldfish in a bowl of water. Obviously, the goldfish does not immediately go to thermal

equilibrium like our cream did. In fact, the goldfish would remain in the state he is in until he died of starvation, or continued to live if food was available. This in physical nature is not what occurs typically in the universe, hence, Schrödinger's definition. So the question persists: How does life defy the second law of thermodynamics?

Some have stated the sun is responsible for life on earth, and hence responsible for this apparent contradiction to the second law. Bill Nye (The Science Guy) used this argument in a recent debate with Ken Ham. This appears to make sense intuitively, but is not fully accurate. If the sun were to continue to shine 24 hours a day on the earth, life could not exist. In fact, it would never start. What matters is not the energy from the sun but that the energy is at low entropy. Without the rotation of the earth and the days being divided between light and dark, life would be impossible. Actually, the earth gives off more energy than we get from the sun through radiation that occurs at night as infrared light. Physicists have quantified this; for every one photon from the sun, the earth gives back 20 photons to the universe at night. We increase entropy by a factor of 20. From a physicist's standpoint, the global nature of this argument is that life on earth does not violate the second law of thermodynamics.

As we have already stated, one quality that defines life is the ability to replicate itself with the most popular theory to date being that life began in an RNA world. However, some physicists believe in a metabolism-first theory regarding the origins of life, not the RNA world. They argue complex chemical reactions take advantage of the low entropy environment on the earth (sort of hardware before the software approach). But this creates a problem in and of itself. Going from an original situation of low entropy to high entropy requires a chain reaction that is very difficult to get started! The argument is that reactions occurred in very specific geologic formations; but do you see a problem with this argument? Specific geologic formations would imply design which would require information, the source of which cannot be explained by physicists.

Physicists then postulate that life started on earth in order to increase the entropy of the primitive earth's atmosphere. To me this is somewhat circular. Why would the earth want to increase its entropy and who would direct it to form life to do so? The bottom line is, physicists know how difficult it is to explain the origins of life; therefore they usually stay out of that argument. They would sooner describe relativity theory and quantum mechanics, which they find much, much easier to do and they know it!

Our present universe, from a physical standpoint, simply does not give the answers to physicists on how life began. In order to explain it they invoke the many worlds hypothesis, sometimes referred to as the multiverse hypothesis. In this case, an infinite number of universes can be produced that would eventually generate life. In other words, sooner or later some universe had to acquire life-sustaining characteristics and if there was an infinite number of parallel universes this could occur. To me, this is sort of parallel to the evolutionary argument. Given enough time, evolution could occur. From a physicist's standpoint, given enough universes, life could occur. Stephen Meyer has stated, "That some scientists dignify the many worlds hypothesis with serious discussion may speak more to an unimpeachable commitment to naturalistic philosophy than to any compelling merit for the idea itself." Amen to that!

The proof of entropy is all around us in actuality. We understand that disorder increases. Cars rust and machines wear out and there is no spontaneous reversal of this process. This applies to living systems as well. A dead plant has no information within it to convert the sun's energy to useful work. Death is the ultimate consequence of entropy. Andrew McIntosh, Ph.D. in Mathematics from the University of Wales and the Cranfield Institute of Technology, has concluded, "What is paramount in this discussion is the information (that is the setting of rules, the language, code, etc.) has been there from the beginning." The physical laws of the universe do not explain this. As McIntosh observes, chemicals do not define the message they are carrying and information cannot be defined simply by physics or chemistry. Ker C. Thomson, geophysicist, clearly and correctly states that only a purposive machine could "temporarily" obviate the second law's demands. "The idea that the second law can be confined to a closed system is a piece of confusion on the part of the proponent of such concept." Dr. Thomson goes on to say, "The second law tells us clearly that life could never get started by the activities of matter and energy unaided by outside intelligence."

Going back to our original question: Does life violate the second law of thermodynamics? From a physicist's standpoint, it requires them to invoke theories or postulates which they know are not provable in the laboratory setting. When one has to use multi-universe theories or defer to explaining life's apparent contradiction of the second law as being the result of a "non-closed system" and only temporarily decreasing entropy for the wider purpose of increasing it, or possibly the biggest copout of all merely stating that life exists to increase the overall entropy of the universe, they are relying on poor arguments and defy the obvious, that life is truly the result of a grand design from a higher intelligence. John Cimbala, Ph.D. in Aeronautics from the California Institute of Technology, has stated, "Only someone or something not bound by the second law of thermodynamics could be responsible for the universe. Only the creator of the second law of thermodynamics could violate it and create energy in a state of availability in the first place." Dr. Cimbala concludes that "the universe had a beginning, and that beginning had to have been caused by someone or something operating outside the known laws of thermodynamics."

Chapter 13

Evolution and Racism

Marvin Lubenow in his book *Bones of Contention* raises the question: Is evolution racist? Actually that question has been asked by many in the past. By its very nature evolution lends itself to racism. As a result, evolution at the very least has been responsible for many atrocities in the history of man including the dissolution of several cultures and even the holocaust that occurred in World War II.

Clearly, Darwin was a racist, as were most of his fellow Englishman as well as a large percentage of the French and Germans. In fact, the full title of Darwin's famous book is *The Origin of Species or the Preservation of Favored Races in the Struggle for Life*. Although the term race and species were used interchangeably by Darwin it is clear that Darwin felt the processes of evolution preserved and favored certain races.

Racism is defined as a perceived inherent superiority of certain races which invokes prejudice and hatred for those races said to be inferior. Although racism has been practiced throughout human history it certainly is not condoned from a biblical perspective. The Bible clearly teaches that mankind was of one blood and the second most important precept of Christianity, that is treating your neighbor as yourself, precludes racism. And although evolution is not the cause of racism as Marvin Lubenow points out, the sin of evolution is that it gives humans an allegedly scientific justification for it. Various groups have been denied full humanity throughout history including women, slaves, various races, the mentally and physically disadvantaged, the aged, the poor, and in some cases children. As a result of evolution, past races were denied full humanity as we have already seen including the Neanderthals and other cultures. One can even say that the concept of evolution is behind the racism of abortion.

When Darwin began his voyage on the *Beagle* at the age of 22 in December of 1831, there were on board that vessel three "savages" from Tierra del Fuego which is at the tip of South America. These individuals had originally been captured by Robert Fitzroy of the *Beagle* on a previous voyage and were taken to England by Fitzroy in order to civilize them, make Christians of them, and return them as missionaries. Because of the degree of culture the Fuegians had acquired while in England, Darwin was unprepared and shocked when encountering the Fuegians in their native habitat. Darwin wrote, "It was without exception the most curious and interesting spectacle I have ever beheld. I could not have believed how wide was the difference between savage and civilized man; it is greater than between a wild and domesticated animal, in as much as in man there is greater power of improvement." Darwin would go on to point out after viewing them, "One can hardly make oneself believe that they are fellow creatures, and inhabitants of the same world."

One of the more unfortunate incidences in human history is the almost complete genocide of the Tasmanian aboriginals in the 1800's. The Tasmanian aboriginals were black people who lived in a tiny island to the south of Australia and isolated there for apparently thousands of years. Europeans invading their tiny island considered them subhuman and justified this from an evolutionary perspective. When the Europeans first arrived in Tasmania there were approximately 5,000 aboriginals. By 1842, only 135 remained. Darwin in his book *The Descent of Man* seems to justify this. He states, "Extinction follows chiefly from the competition of tribe with tribe, and race with race. When civilized nations come into contact with Barbarians the struggle is short. We can see that the cultivation of the land will be fatal in many ways to savages, for they cannot, or will not, change their habits." Although the genocide of the Tasmanians occurred before the publication of Darwin's book, the concept of evolution was there long before it. Today there are no full-blooded Tasmanians but only those of a mixed race estimated at about 4,000.

One can also make an argument that slavery as practiced in Europe in the 16th, 17th, and 18th centuries and in The United States was a product of racism, but it was also a product of pre-evolutionary thought. The idea that one race has the authority to subjugate another race because of the feeling that the enslaved people are in some way inferior and even some cases subhuman epitomizes racism. How else can one justify the tens of thousands of Africans who were rounded up like animals, taken away from their home land against their will, shackled and bound, and sent to foreign lands, many of whom died on the way, and once in those lands treated as common property, separated from their families, bred like animals, and in many cases treated in the most inhumane manner.

Some, in the name of Christianity, have even tried to condone such practices. To say slavery as practiced in Europe and America in the 19th century has a scriptural basis, shows either complete ignorance of the Scripture or ignorance of how slavery was actually practiced. Scriptures clearly teach that all humans are created in his image. In Genesis 1:26 God said, "Let us make man in our image, and our likeness...." In Exodus 21:16, the Hebrews were to put to death anyone who kidnapped a man "whether he sells him or he is found in his possession." Further, all humans constitute one large interrelated family as noted by Luke in Acts 17:26, "From one man He made every nation, that they should inhabit the whole earth." As we have already alluded to, the "golden rule," as stated by Christ is to treat one's neighbor as oneself (Matt. 7:12). Kidnapping, murder, torture, and human enslavement clearly are not treating one's neighbor as oneself. Finally, God has intended for all mankind to be evangelized and the apostles were given "all authority in heaven and in Earth" and were to "go and make disciples of all nations" (Matt. 28:18-19). All of humanity constitutes these nations.

Since the twentieth century, one of the most egregious examples of man's inhumanity to man was the holocaust perpetrated on Jews, Gypsies, nonconformists, and the mentally and physically disadvantaged that occurred in Nazi Germany and Eastern Europe during World War II by Adolf Hitler and his minions. Hitler's hangman at the time was Adolf Eichmann. Eichmann was personally responsible either directly or indirectly for the deaths of over 5 million Jews. After the war, the State of Israel eventually discovered Eichmann in 1960 hiding out in Argentina after which he was brought back to Israel to stand trial for war crimes against humanity. In *Bones of Contention*, the late Dr. A.E. Wilder-Smith relates an incident concerning Eichmann after he had been convicted of these war crimes. He was sentenced to hang and before his hanging a chaplain had asked if he would like to make a confession. Eichmann replied, "Confess? What have I got to confess? I've done nothing wrong!" The chaplain then replied, "You've done nothing wrong? Do I understand you?" "Yes,"

Adolf Eichmann

Eichmann replied, "I've done only right!" When the chaplain asked for him to explain himself Eichmann stated, "Both the churches in Germany, the Catholic and the Protestant, believe in theistic evolution, both of them believe that God's method of creation was to wipe out the handicapped and wipe out the less-fitted. And as the Jews are less fitted than our people, I have only helped God in His methods. I have only catalyzed God's way of working. And when I meet God I shall tell Him so."

Recently on a trip to Normandy, France, I was able to visit the war museum at Caen. The atrocities of Germany were well documented at the

museum but one particular poster that caught my eye was that of an obvious mentally and physically handicapped man sitting in a chair. Behind him were two nurses and to the side appeared to be a physician. The heading on the poster in German translated into English was, "Can we afford to continue to take care of people like this?" The poster went on to document how much money the German state was spending to keep this man alive and pointed out to the citizens that it was in essence their money that was doing so. The implication of this is obvious. What do you think could be the justification of the German people for behavior like this? Pure and simple it's evolution.

A poster advocating that handicapped people be eliminated.

As Lubenow points out, Hitler was not a mad man but he certainly was a very bad man. He clearly believed the German people were genetically and inherently superior to the Jewish people. The gas chambers of Auschwitz and other concentration camps were Hitler's way of practicing "natural selection" and the survival of the fittest.

Racism and genocide are certainly not unique to Germany as we have

already seen concerning the Tasmanian aborigines. The late Steven J. Gould, himself an evolutionist, recounts a story of Dutch settlers who shot and ate African Bushmen. They believed the Bushmen represented a race inferior to them and saw no difference between them and Orangutans or any other animals. Gould points out, "The tragedy of this is that the history of scientific views, for the most part and until quite recently on the subject of race, have supported this social and political notion that human groups are separated by profound and innate inequalities in intellectual abilities and moral behaviors."

In the 1960's, Boston born and Harvard-educated Anthropologist, Carlton S. Coon, created somewhat of a firestorm when he postulated the concept of five basic human races. The five races he proposed were: 1) Caucasoid (a group mostly European and North African as well as near Easterners and peoples of India and Pakistan), 2) Mongoloids (a group that includes East and Southeast Asia and Indonesia as well as the Polynesians and American Indians), 3) Australoids (a group that includes natives of Australia, New Guinea, and Melanesia as well as aboriginal tribes of India and the negroid dwarfs of Indonesia), 4) Capoids (a group that included Bushmen and Hottentots) and 5) Congoids (a group that includes African pygmies and negros). Coon believed each of these races developed separately into *Homo sapiens* and their stock could be traced back to *Homo erectus*. Coon was apparently not a racist himself and tried to avoid racial overtones and used "scientific evidence," nonetheless, his beliefs created quite a stir in the anthropologic and evolutionary circles. Although his concepts were discredited, they are somewhat similar to a current theory of human evolution known as the multiregional continuity model. To escape the charge of racism, paleontologists have now derived a model in which "1) All living humans evolved from the very same human stock; 2) All evolved in the same period; and 3) Thus the racial distinctions among humans today have a very shallow rather than a very deep history," says Lubenow. As a result of this what we now have is called the "Out of Africa Model." Because of this, the role of Neanderthal man has been downgraded and he is no longer considered a precursor to modern man. As you recall, Neanderthal fossils were found in Europe and to give Neanderthal pre-eminence would now have paleontologists appear to be racists.

The Out of Africa Model, the politically correct model for evolutionists, began in 1987. Through study of the DNA as well as fossil records of African Eve this Out of African Eve Model now postulates, 1) *Homo erectus* or Archaic humans originally evolved in Africa from *Homo habilis* or Australopithecine stock. 2) Modern humans evolved in Africa stemming from Mitochrondrial-Eve who lived in Africa 200,000 years ago. 3) Modern humans migrated out of Africa about 100,000 to 150,000 years ago. 4) Modern humans then migrated to Europe and into Asia. 5) These modern humans eliminated all other humans with little or no interbreeding replacing Neanderthals in Europe and all other "primitives" in the world. 6) Africa thus became the birthplace of all humans. Of course, the validity of this view cannot be established by the fossil record.

In 1987, Berkeley biochemists examined the mitochondrial DNA (mtDNA) of 136 women from all over the world and from varying races. Assuming that all changes in mtDNA were the results of mutations over time and mutations occur at a constant rate, they concluded that the original "mother" of all humanity lived in sub-Saharan Africa 200,000 years ago. This was all computer generated and as we have seen in our study computer generated models have their own little problems. In this case, the computer was designed to reveal a maximum parsimony phylogeny (that is, only a limited number of family trees). The computer was given information and it was thus biased by the order in which the data was entered. Some such as Henry Gee on the editorial staff of *Nature* described their results as "garbage." Mark Stoneking of the Max Planck Institute for Evolutionary Anthropology in Leipzig acknowledged that the African Eve Model has been invalidated.

The African Eve Model with its computer-generated analysis of mito-chondrial DNA employs circular reasoning. The DNA analysis is based on the assumptions of mutations in the DNA nucleotides. When looking at human DNA nucleotides, how does one know which ones are the results of mutation and which ones have remained unchanged? Marvin Lubenow points out, "He [the evolutionist] uses as his guide the DNA of the chimpan-zee. In other words, the studies that seek to prove that human DNA evolved from chimp DNA start with the assumption that chimp DNA represents the original condition (or close to it) from which human DNA diverged. This is circularity with a vengeance." When evolutionary interpretations of fossils are superimposed on the DNA model the circularity is even more obvious because an evolutionary interpretation of the fossils is its starting point. These sophisticated computer programs are "independently able to generate truth," but in reality they cannot.

Mr. Lubenow states, "At this time, the African fossil evidence for African Eve is both farce and questionable." Jonathan Marks of Yale University notes, "And with each new genetic study that claims to validate Eve conclusively there comes an equal and opposite reaction showing the study's weakness."

I believe it is obvious that evolution is racist at its basic core. Of course I am not saying that present evolutionists are racist by any means. In fact, they are making every effort to separate themselves from the past. That is why evolution and genetics are now being used in an attempt to prove human equality via the Out of Africa Model. It is, however, somewhat odd that the most racist theory ever to gain acceptance in the scientific community is now being given credit for proving human equality. We know as Christians this is unnecessary; human equality does not require proof. It is intrinsic to God's creation.

Chapter 14

Alternative Theories

For Neo-Darwinism to explain evolutionary changes that produce new species three fundamental processes must occur. First, change must be the result of random minute variations or mutations. Secondly, the process of natural selection takes these changes and variations and mutations and selects out the traits that will allow for the survival of the most fit. And thirdly, these variations must be heritable. As we have seen in the course of this study, Neo-Darwinism has significant flaws and hence is extremely unlikely to produce changes in organisms that would result in new kinds.

Many evolutionary biologists have serious doubts of orthodox Neo-Darwinian theory and as a result, new theories have been postulated concerning evolution. The biggest problem that evolutionary biologists now have with Neo-Darwinism has to do with gradualism; that is the immense amount of time required to produce significant change resulting in a new morphologically distinct organism. The common denominators with all these new theories are their attempts to speed up the process or to find mechanisms that would allow for this.

These evolutionary biologists are keenly aware that the fossil record simply does not support Neo-Darwinism. This is especially obvious when they look at the Cambrian geological layer. Simon Conway Morris, a renowned Cambrian Paleontologist, understands the dilemma posed by the Cambrian Explosion. He stated in 2009 in an essay published in *Current Biology*, "Everywhere else in the *Origins* the arguments slide by skillfully into place, the towering edifice rises, and the Creationists are left permanently in the shadow." But not when it comes to the seemingly abrupt appearance of animal fossils. For Conway, this "opened the way to a post Darwinian world." As a result, new theories now abound. To be sure, these new theories are still "evolutionary and naturalistic"; they either attempt to modify Neo-Darwinism or offer completely new mechanisms.

Historical Background

Since the publication of Darwin's book, *On the Origin of Species*, scientists have always questioned its scientific merit as we have seen earlier. In the 1930's and 1940's, Richard Goldschmidt postulated that radical transformations were responsible for evolution. He seemed to endorse Otto Schindewolf's concept that, "The first bird hatched from a reptilian egg." In other words, for evolution to have occurred, giant leaps in morphological variaton were necessary. Macro mutations would be necessary and Goldschmidt proposed a "hopeful monster" that could explain sudden onset of distinct species which, of course, is what the fossil evidence seemed to indicate. Scientists did not seriously consider Goldschmidt's proposal, yet in 1972 Niles Eldredge proposed the concept of punctuated equilibrium. Later, evolutionist, Stephen Jay Gould, would become the chief spokesman for punctuated equilibrium or "punk eek." Gould and Eldridge found the absence of changes in the fossil record very interesting, therefore, proposing a mechanism known as "allopatric speciation." Briefly, allopatric speciation proposed the generation of new species from a separate parent population with selection based on species competing with each other. They drew on insight from population genetics to explain new genetic traits. The problem is, population genetics teaches that in typically large populations of organisms it is quite difficult for new genetic traits to spread throughout an entire population. Most biologists by this time believe that allopatric speciation requires too much change too quickly to provide the theory of punctuated equilibrium with a biologically viable mechanism for producing new traits or morphologic forms of animal life. Stephen Meyer writes, "Few if any evolutionary biologists now regard punctuated equilibrium as a solution to the problem of the origin of biological form and novelty." To the credit of Eldredge and Gould, punk eek at least made an attempt to explain the evidence presented by the fossil record albeit a feeble and seriously flawed effort.

Evo-Devo

Evolutionists continue to search for a mechanism that is both viable and fits the current evidence. In the latter part of the 20th century, evolutionary biologists such as Rudolf Raff, Sean B. Carroll, and Wallace Arthur developed a discipline of biology known as Evolutionary Developmental Biology, also referred as "Evo-Devo." There are multiple variations of Evo-Devo but they all challenge a key aspect of Neo-Darwinism; that is, instead of new forms arising as a result of slow incremental accumulation of minor mutations, Evo-Devo argues that mutations affecting genes involved in animal development can cause large-scale morphologic changes and even change new body plans.

Evo-Devo involves at least three distinct approaches. One approach has mutations producing modifications in large increments. As you can see, this is very similar to punctuated equilibrium proposed by Gould and Eldridge. The second Evo-Devo approach is Neo-Lamarckism; that is epigenetic inheritance. In this mechanism, inheritable alterations are passed on through epigenetic information. The final Evo-Devo approach is referred to as natural genetic engineering. Natural genetic engineering affirms non-random genetic rearrangements drive evolution.

The problem with all Evo-Devo mechanisms is they contradict the results of 100 years of mutagenic experiments. Early acting developmental muta-tions have never been able to cause stable large-scale changes in animal body plans. As we have previously seen in our study, major changes produced by mutations are not viable and viable changes are not major. In either case, these kinds of mutations could never produce the major changes necessary to build new body plans.

In an attempt to circumvent this, many Evo-Devo biologists believe the Hox genes are responsible for macroevolutionary changes. You may recall that the Hox genes are those genes that regulate the expression of protein coding genes in the process of animal development, thus they are similar to genetic switches rather than genetic blueprints. The problem for the Evo-Devo biologists is that because the Hox genes coordinate the expression of so many other genes, mutations in the Hox genes have proven universally harmful. By altering the structure of the animal you are producing a non-viable organism. Also, Hox genes are expressed after the beginning of animal development and hence well after the body plan has begun to be established. Hox genes, therefore, don't determine body plan formation. Thirdly, Hox genes provide information for building proteins that function as switches and turn off or on other genes. The Hox genes themselves do not contain information for building body structures. In fact, epigenetic information in structures actually determines the function of many of the Hox genes.

Another theory from the evolutionary development camp comes from Michael Lynch, geneticist at Indiana University. Lynch has proposed a neutral, or "nonadaptive" theory of evolution. Without going in great detail, Lynch summarizes his theory by stating, "Three factors (low population size, low recombination rates, and high mutation rates) conspire to reduce the efficiency of natural selection with increasing organism size." By stating this, Lynch has advanced a powerful mathematical critique of the efficiency of Neo-Darwinian mechanisms. He believes in small population's "neutral processes" such as random mutation and genetic recombination and genetic drift could pre-dominate their effects over natural selection. Lynch's theory unfortunately fails in that it offers no explanation for some of the crucial molecular machinery needed in the eukaryotic cell to produce changes in form. He also overlooks crucial epigenetic information and the fact that genetic drift, that is the change in frequency of a gene variant in a population due to random sampling of organisms, does not favor beneficial mutations. Finally, Lynch vastly underestimates the waiting times required to generate complex adaptations and his mathematical computations consider only those special paths that lead directly to the desired end.

Yet another Evo-Devo theory involves Neo-Lamarckian epigenetic inheritance. You may recall Jean-Baptiste Lamarck believed in the concept of inheritable acquired characteristics. With the discovery of Mendel's laws in the 1900's this was essentially put to rest; however, there are some modern Neo-Lamarckians who think that non-genetic sources of information and structure may place some role in the evolution of biologic forms. Eva Jablonka of Tel Aviv University believes that Neo-Lamarckism "allows evolutionary possibilities denied by the 'modern synthesis' version of evolutionary theory, which states that variations are blind, are genetic (nucleic acid-based), and that saltational events do not significantly contribute to evolutionary change." Neo-Lamarckian mechanisms cannot, however, explain the origin of animal form. Macroevolution requires stable, that is permanently heritable changes, but Neo-Lamarckian mechanisms involve structures that either do not change or do not persist over more than several generations.

The final Evo-Devo theory proposed for evolution comes from James Shapiro of the University of Chicago and has been called "natural genetic engineering." He observed that within populations organisms often modify themselves in response to different environmental challenges. Shapiro favors an evolutionary view that emphasizes pre-programmed adaptive capacity or engineered changes. In his model, organisms could respond in a "cognitive way" to environmental influences changing or mutating their own genetic information. The obvious flaw in Shapiro's model concerns information.

Where does the programming come from that accounts for all these adaptive capacities? Without the concept of an intelligent designer there is no answer to this question.

Self-Organization

The final hypothesis we will take a look at is referred to as the self-organizational model proposed by Stuart Kauffman in 1993 in his treatise, *The Origins of Order: Self-Organization and Selection in Evolution.* Steven Meyer in his book, *Darwin's Doubt,* notes that Kauffman proposed, "First, that gene regulatory networks in animal cells (genes that regulate other genes) influence cell differentiation. They do this by generating predictable 'pathways' of differentiation, patterns

Stuart Kauffman

by which one type of cell will emerge from another over the course of embryological development as cells divide." To put it another way, networks of regulatory genes in embryonic cells determine the pathway by which cells divide and differentiate. Kauffman suggested a self-ordering property inherent in a wide class of organisms. For evolution to occur "long-jump mutations" must occur according to Kauffman. As Meyer points out, self-organizational processes cannot account for the origin of animal body plans such as those seen in the Cambrian explosion. This theory presupposes the existence of a "developmental genetic toolkit," that is "a whole set of genes, including regulatory genes, that help to direct the development of animal body plans." We are back to the same problem that all these theories have and that is: Where does the information come from to direct these changes?

These self-organization theorists have cited crystals, vortices and convection currents as evidence for the supposed power of physical processes having the ability to generate "order for free." The problem is, the type of order seen in these physical occurrences in nature have nothing to do with specific order of arrangement. The information contained in DNA, RNA and proteins on the other hand are characterized not by simply order or complexity but instead "specified complexity," a concept elucidated by William Dembski. Life cannot be described or explained simply by referring to natural law or law-like self-organizational processes. As we have seen, the sequences of nucleotide bases and RNA cannot be attributed to self-organization forces, the origin of these specific arrangements must come from somewhere else. The same can be said also for epigenetic information, the origins of which can also not be attributed to self-organization forces. Crystals, geologic figures, pattern lines, triangles, vortices, and spirals may be self-organized but they do not exhibit specified complexity characteristic of structures such as DNA and RNA found in life. Meyer points out from a meeting he had in 2007 with evolutionary biologists, that some of them even admit, "Self-organization is really more of a slogan than a theory" and even Kauffman himself calls self-organization "natural magic." In an MIT lecture he concluded, "Life bubbles forth from natural magic beyond the confines of entailing law, beyond mathematization."

Since the beginning of modern science, scientists have always held to the chief scientific reasoning known as *vera causa*. This principle in essence holds that events or phenomena require an identifying "true cause" identified through experimentation. Clearly, self-organizational theories have failed to provide a *vera causa*. Again from Stephen Meyer, "Instead, they either beg the question as to the ultimate origin of biological information or point to the physical and chemical processes that do not produce the specified complexity that characterizes actual animals."

Conclusion

For evolutionary biologists the mystery still remains. It is unfortunate that the answer to this mystery is so painfully obvious and were it not for their naturalistic views could be readily appreciated, that is intelligent design. Christians know that God, the all omniscient, is the source for this intelligence. Intelligent design challenges the idea that natural selection and random mutations (and other similarly undirected materialistic processes) can explain the most striking appearances designed in living organisms. The only two possibilities that I am aware of is either life is the result of purely undirected material processes or something or someone of intelligence played a role. To me, the evidence is overwhelming. Peter Lipton, a contemporary philosopher of science, has stated, "The best explanation is the one that best explains the facts or that best explains the most facts." Clearly, the best explanation is that there is an intelligent designer, God, behind the origins of life.

Chapter 15

The Elephant in the Room

After examining all the evidence and looking at evolution from a purely scientific point of view, I believe there is little scientific evidence to support it, no matter how dogmatic its leading proponents are. We have seen that the mathematical probability of life springing forth spontaneously by purely naturalistic forces is virtually zero. We have also seen that natural selection cannot possibly be  a valid mechanism for change in kinds and there is a significant paucity in the fossil evidence to support the evolutionary thesis. No doubt evolution has a certain philosophical elegance but elegance is not scientific.

This is a problem quite frankly that I have struggled with and that is, if evolution is so obviously philosophical and so diametrically opposed to the scientific method how can so many very smart, extremely well-educated people, accept it so easily and in most cases without question? Am I the only one who can see the fallacy of evolution? Of course the answer is no. Many, many scientists today and in the past have rejected evolution on a scientific basis, some of which we have already noted in this book. People like Danny Faulkner, Astrophysicist; William Dembski, Mathematician; Robert Kaita, Physicist; Stephen Meyer, Scientific Philosopher; Michael Behe, Biochemist; David Berlinski, Mathematician; William Bradley, Mechanical Engineer; Nancy Darrall, Botanist; Angela Meyer, Horticulturist; Henry Zull, Biologist; John Kramer, Biochemist; Elaine Kennedy, Geologist; John Morris, Geological Engineer; Robert Collins, Physicist; William Lane Craig, Philosopher; do not accept molecules-to-man evolution and this list could go on and on and on. So the question remains, why can some see the problems scientifically with evolutionary theory and others cannot? This seems to me to be the elephant in the room!

To answer this question we need to first examine why people accept evolution in the first place. Almost a half a century ago Bolton Davidheiser in his book, *Evolution and Christian Faith*, divided evolutionists into groups and I believe his ideas are still cogent for answering our primary question.

Most scientists today that accept evolution fall in the first category we will discuss and they are those who believe evolution because they are convinced it is true and simply cannot take any other view. Considering the evidence is the same for everyone why is this so? First and foremost is the fact that evolution in its various permutations is the only naturalistic theory currently put forth that explains the origins of life and many cannot accept any other view, especially if it implies a supernatural cause. Since they cannot accept an intelligent design explanation they are blinded by evolution and cannot see the evidence against it. It is as if they are colorblind, they cannot accept any other view. Take for example the picture on this page, what do you see? Some readily see a duck but for others it is obviously a rabbit yet the picture is the same for both. When one eliminates intelligent design or creation from the argument, evolution becomes the logical explanation for life. But when one accepts the possibility of an intelligent cause the overwhelming evidence supports it and even evolutionists admit this.

Consider a few examples to prove the point. Dr. M.S. Watson, the late Zoologist of London University, has written, "Evolution is a theory universally accepted, not because it can be proved true, but because the only alternative, special creation, is clearly impossible." Richard Lewontin, a Harvard geneticist, in the 1997 New York Review of Books opined, "We take the side of science in spite of the patent absurdity of some of the constructs. In spite of its failure to fulfill many of its extravagant promises of health and life, in spite of the tolerance of the scientific community for unsubstantiated just-so stories, because we have a prior commitment, a commitment to materialism. It is not that the methods and intuitions of science somehow compel us to accept a natural explanation of the phenomenal world, but, on the contrary, we are forced by our prior adherence to natural causes to create an apparatus of investigation and a set of concepts that produce material explanations, no matter how mystifying to the uninitiated. Moreover, that materialism is absolute, but we cannot allow a divine foot in the door."

Even the courts of our country have sided with materialism. Consider the opinion given by Judge William R. Overton in a famous court decision he rendered in Little Rock, Arkansas that prohibited the mention of creationism in the classroom. In his 1981 court brief, he defined science as, "1) It is guided by natural law; 2) It has to be explanatory by reference to natural law; 3) It is testable against the empirical world; 4) Its conclusions are tentative, i.e. are not necessarily the final word, and 5) It is falsifiable." But is this really what science is? In the first two points Judge Overton is not defining science but naturalism and they are not the same. Think about it, if Overton's definition of science is correct then most of the scientists in the past including men such as Isaac Newton and James Maxwell and many, many others, were not practicing science since they accepted a divine cause. Evolution is not science even as defined by Judge Overton, who incidentally is neither a scientist, a philosopher of science, or a science historian, but it is naturalistic and that is why many scientists accept it. J.P. Moreland has stated it very well, "People who fought creationists for deriving their scientific hypothesis from The Bible or a theological framework commit the generic fallacy and are out of touch with the actual way science has been practiced repeatedly throughout history."

Geoff Downes, in the book, *In Six Days*, illustrates the absurdity of naturalism very well. Say for example, a person finds a dead body in the park with a knife stuck in its back. The most logical explanation would be to assume the person died from some outside agency, but if we believe natural causes are the only way someone can die we will never arrive at the correct conclusion; that is a murder most likely took place. It becomes now a matter of assumption and not science, just like evolution. To be fair, the vast majority of scientists who accept evolution have never taken the time to examine the hard evidence or lack thereof for evolution because it is not in their realm of specialty. They leave this to their fellow scientists, paleontologists or evolutionary scientists, because to believe otherwise is not naturalistic. What they may not understand, however, is that evolutionary scientists employ a certain logic that they probably would not agree with.

That logic is the abductive inference that Stephen Meyer's in *Darwin's Doubt* eloquently elucidates. The deductive argument, which is the argument used by the pure, natural scientists, goes like this. The major premise is this; if A has occurred then B will follow as a matter of course. The minor premise is A has occurred. The conclusion being then B will follow as well. Meyer gave this as an example. Major premise: all men are mortal. The minor premise: Socrates is a man. The conclusion: Socrates is mortal, i.e., he will die.

Abductive reasoning is somewhat different. The major premise in abductive reasoning is this, if A occurs then B would be expected as a matter of course. The minor premise, the surprising fact B is observed. Hence, the conclusion there is reason to suspect A has occurred. This is another example given by Meyer. Major premise: a mud slide occurred; we would expect to find felled trees. Minor premise, we find evidence of felled trees. Therefore, the conclusion is we have reason to think a mud slide may have occurred. Using our previous story of Geoff Downes, the major premise men die of natural causes, the minor premise we find a man dead in the forest, the conclusion therefore is the man died of natural causes. As you can see, there is an obvious fallacy in abductive reasoning, that being the failure to acknowledge that more than one cause, or antecedent, might produce the same evidence (or consequent). Great scientists of the past such as Newton, Maxwell, Einstein and Feynman would never use abductive reasoning, yet this is precisely the reasoning used by the evolutionary scientists.

So is it possible to persuade a naturalist to consider intelligent design or creation? It is certainly very difficult but there are multiple examples of scientists that have done that very thing. The book, *In Six Days*, edited by John F. Ashton, Ph.D., gives multiple examples of scientists who began their careers as naturalists but now have turned to intelligent design and creation as a better explanation for the beginnings of life and even the beginnings of the universe. Scientists such as James S. Allan, Geneticist from the University of Edinburg, John M. Cimbala, Ph.D. in Aeronautics from California Institute of Technology, A.J. Monty White, Chemist, and Colin W.P. Mitchell, Geologist with a Ph.D. from Cambridge University, are just a few examples of scientists who were once atheists or agnostics who now believe in a divine creator. One can then change his point of view but admittedly this is very, very difficult.

Most scientists I would say are not necessarily antireligious. It simply doesn't fit their paradigm. They would not attempt to persuade or dissuade someone from his religious beliefs, it is just not what they accept. There is, however, a second group that Davidheiser alludes to and that is evolutionists whose goal seems to be to destroy religion in general and especially Christianity. Any group or people that believe in a divine creator this group would vehemently oppose. I would place in this category those who belong to the National Center for Scientific Education headed currently by Eugenie Scott. This group attacks school districts and states who would dare allow intelligent design to be included in any biology textbook. Many times the NCSE has called on the American Civil Liberties Union (ACLU), who threatens school boards with expensive lawsuits if creation is even mentioned in class or in a textbook. In some cases, the ACLU has even gone after individual teachers for trying to incorporate intelligent design in their class with the pretext being that creation is not scientific and religion is superstition. They hide behind the separation of the church and state argument and to date their tactics have been very successful but they are not above distorting the facts to achieve the agenda.

More recently, the NCSE has changed its strategy a bit. Ken Ham notes in his book, *The New Answers Book 4*, that in 2008 Ms. Scott on behalf of the NCSE posted on her website an article entitled, "How you can support evolution education." One section in that article listed three ideas. 1) Institute adult religious projects to focus on evolution with the religious leaders. 2) Encourage religious leaders to support the "Clergy Letter Project" and to participate in evolution education. 3) Encourage leaders to produce educational resources about evolution and education to stand for evolution education. Ms. Scott understands she cannot win the argument with science; therefore, she realizes that she and the NCSE must ally themselves with "mainstream clergy." These clergies then go to school boards to support evolution. Second Corinthians 6:14, "Do not be bound together with unbelievers; for what partnership has righteousness with lawlessness, or what fellowship has light with darkness," and Matthew 7:15, "Beware of false prophets who come to you in sheep's clothing but inwardly are ravenous wolves" come to mind. With this group, it would be virtually impossible to persuade them to accept the idea of intelligent design or creation; it would be like casting pearls before swine.

This group would also include the "new atheists" whose agenda is ridding the world of all religion. Chief proponents of this group include Richard Dawkins and theoretical physicist, Lawrence Krauss. In the documentary, *The Unbelievers*, Lawrence Krauss states, "You've got to confront silly beliefs by telling them they are silly. If you're trying to convince people, pointing out what they believe is nonsense is a better way to bring them around." For the new atheists, a recurring theme is evidence of reality. Ostensibly they're not opposed to spirituality but for them spirituality is more awe and wonder rather than a belief in anything supernatural. For the new atheists, atheism is their religion evidenced by the fact they actively proselytize. Anyone who does not come over to their way of thinking is either stupid, insane, or fanatical.

The third group Davidheiser has written about are those who believe in evolution, who know little about it but follow along because they are told it's true and to not believe it would be going against the mainstream. After all no one wants to appear stupid and in not accepting evolution one is likely to feel so. This is really not hard to understand since intelligent design is not allowed to be taught in schools. High school biology teachers, most of whom are certainly not Ph.D. trained, are taught the mantra of evolutionary fact and high school textbooks continue to use the old icons of evolution, all of which are seriously flawed, and students simply are given no other alternatives. You see, if you were told, for example, that "There are hundreds or even thousands of transitional forms and this is verified in textbooks" even though examples other than Eohippus are rarely given, what is a person to believe? This group by far represents the vast majority of people who believe in evolution or at least accept it. I believe people in this category can be persuaded of the truth as long as they have an open mind to do so. We do not have to be scientists to believe and I think it is an excuse to say only "experts" are qualified to decide which belief is correct. If the arguments in this book are seriously considered then a person can make an appropriate decision and see the facts for what they are. People can change and if that were not so, where would any of us be?

Section 2

An Alternate View

Preface

If naturalism and evolution are not the most plausible explanation of the origins of the universe and life, then what is? How about a creator! I believe a creator or an intelligent designer is not only the more logical choice, but also better fits the scientific evidence. If there is a creator, what characteristics would you expect him to have? I would expect a creator to be omniscient, omnipresent, and transcend time, space, and natural law, wouldn't you? His creation would exhibit exquisite design and demonstrate supreme intelligence, and everything would fit perfectly. He would set "laws" in motion: laws that would demonstrate a consistency throughout the universe, laws that would also be explanatory, and laws mankind would be able to uncover over time. This fits the model of the Biblical God we know and worship, but would not necessarily be the characteristics intrinsic in the pagan gods of history, many of which have more human-like and fallible qualities.

Henceforth, this book will look at the evidences for a creator by examining how fine-tuned the universe is and how unlikely it could have occurred by natural forces alone. First, however, we will examine the various cosmologies, other than naturalism, that try to explain the beginnings of everything. After that, we will examine the book of Genesis and see if it gives a plausible explanation that fits the facts as we know them. For me, it is only logical that if there is a Supreme Being who created everything, He would communicate this to His ultimate creation, man. Then we will try to determine if it is possible to accurately give an age for the universe or for that matter the earth. Finally, I will try to make sense of all this in our concluding chapter.

Chapter 16

Cosmologies

Before we get into discussing specific evidences, let us look first at the prevailing cosmologies of today. This will make further discussion much clearer. It is not in our scope to study all past cosmologies and current cosmologies, but we will concentrate on those that are the most popular and most relevant to our discussion.

The word "cosmology" comes from the Greek words: "cosmos" and "logos," which mean: "world" and "word." The word *logos* has been generalized to mean "study of" while *cosmos* has been generally understood to mean "universe"; therefore the word "cosmology" means "the study of the universe" as a whole. A related word is "cosmogony," which means the study of the history of the universe, but frequently, these words are used interchangeably.

Two ancient cosmologies that prevailed for centuries included the concept of a flat earth and something called a geocentric cosmology, which implied that the earth was the center of the solar system and the sun revolved around the earth, as did the other planets. Flat earth cosmology was eventually put to rest at about the time of Christopher Columbus, five centuries ago, although the concept of a spherical earth was held by more knowledgeable people at least two millennia before Columbus. The geocentric cosmology has been replaced by a heliocentric cosmology, which now has the sun as the center of the solar system with the planets such as Mercury, Venus, Earth, Mars, Jupiter, and Saturn revolving around it. Since the time of Nicolaus Copernicus in the 16th century and Galileo Galilei, the heliocentric cosmology has prevailed.

The various cosmologies that are most popular today can be broadly placed into four groups: naturalism or molecule-to-man evolution; intelligent design; old earth creationism which includes gap theory, progressive creation, framework hypothesis, and theistic evolution; and finally, biblical creation.

We will first discuss naturalism, the cosmology accepted by much of the scientific community today. The basic premise behind naturalism is that everything in the universe and the universe itself, came about by naturalistic mechanisms with no creator or intelligence responsible for it. Everything began at the "Big Bang" though a limited few still espouse an eternal steady-state universe. Most scientists believe the "facts" support a beginning or at least beginnings of a universe, so we will not discuss steady-state theory here.

The idea of a big bang, as it was derisively called by steady-state theorist, Sir Fred Hoyle, actually had its beginnings with Albert Einstein, who, in 1916, noted that his field equations of general relativity predicted an expanding universe. In 1929, Edward Hubble established that velocities of galaxies result from a general expansion of the universe, and in 1925, Abbé Georges Lemaître, an astrophysicist and Jesuit priest, promoted a big bang creation event. Then, in 1964, Arno Penzias and Robert Wilson discovered 3K cosmic background radiation, which seemed to give evidence to a Big Bang event, eventually winning them the 1978 Nobel Prize in Physics for their work. Finally, in 1992, a team of astrophysicists from the United States reported findings from the Cosmic Background Explorer Satellite (COBE), which again seemed to confirm the Big Bang. Scientists throughout the world described the COBE findings with superlatives such as "the most exciting thing that happened in my life as a cosmologist," per Carlos Frenk of Britain's Durham University; "the discovery of the century, if not all time" by Stephen Hawking of Cambridge University; and a discovery that

was "unbelievably important" and "the holy grail of cosmology" by Michael Turner, astrophysicist, with the University of Chicago.

What occurred at the Big Bang? For this answer, I will look at Stephen Hawking's description in his book, *A Briefer History of Time*, and summarize its postulates. Dr. Hawking

Stephen Hawking

states that what happened before the Big Bang "can have no consequences and so should not be part of a scientific model of the universe." He is in essence, with this statement, ignoring God or anything that occurred prior to a Big Bang model. In the Big Bang model of cosmology, the Big Bang was the beginning of time. With the beginning of the Big Bang, the universe was infinitely small and infinitely hot and all matter that now exists in the universe was compressed into a tiny singularity that may have been about the size of the period at the end of this sentence. When the Big Bang occurred, so the story goes, all elemental particles such as electrons, protons, neutrons, quarks, antiparticles, leptons, baryons, etc., as well as hypothetical particles such as dark matter, came into being in literally nanoseconds. Within one second after the Big Bang, the universe would have "expanded enough to bring its temperature down to about 10 billion degrees Celsius." About 100 seconds after the Big Bang, the temperature would have fallen even more to 1 billion degrees, the temperature inside stars. At this time, protons, electrons, and neutrons would no longer have sufficient energy to overcome the attraction of the strong force; therefore, elemental molecules would have been generated such as hydrogen, which contains one proton and one neutron. This early stage of the universe, which was extremely hot, produced significant radiation and according to George Gamow, produced radiation that would still be around today. This is the microradiation discovered by Penzias and Wilson in 1964 also known as the cosmic background microwave radiation.

A problem does arise, however, in this Big Bang model. The hot Big Bang model does not allow enough time in the early universe for heat to have flowed from one region to another (also known as the "horizon problem" or the particle horizon), according to Dr. Hawking. In fact, Dr. Hawking states, "It would be very difficult to explain why the universe should have begun in just this way, except as the act of a god who intended to create beings like us." To get around this, MIT scientist, Alan Guth suggested the universe might have "gone through a period of very rapid expansion." This has been termed "inflation" and is the only explanation on how the present, smooth, uniform state of the universe "could have evolved from many different non-uniform initial states." Nevertheless, the inflation hypothesis has never been proven, has a series of problems that precludes its viability, and even has many detractors from the secular community.

After the initial state of the Big Bang, within a few hours, the production of helium and other elements would have stopped. Big Bang theorists state that for the next million years or so, nothing much would have happened. At this point, the universe was expanding and cooling, eventually slowed down by gravitational attraction. With the gravitational pull of matter, regions "might start then rotating slightly." This then begins the formation of mass and galaxies. With the formation of galaxies came the formation of planets via the gravitational attraction of cosmic dust, from which eventually, through naturalistic forces, life occurred. The universe, then, according to the Big Bang theory, is 13.8 billion years old and the earth is estimated to be 4.5 billion years old.

"Evidences" for the Big Bang Theory have been nicely summarized by Dr. Danny Faulkner in his book, *Universe by Design*. Dr. Faulker states there are three basic evidences put forth as proof of a big bang event. The first is the presence of cosmic background radiation just discussed, which actually disproves the Big Bang model through the particle horizon problem mentioned above, and the other two are the evidences of an expanding universe and the "abundances of the lite elements" in the universe.

Dr. Faulkner points out, a theory is best judged upon how well it explains data. It should also have some predictive value as well. Dr. Faulkner states, "Many theories have explanatory power but lack predictive power. This is especially true of the historical sciences." What about the three evidences then for the Big Bang? Are they explanatory, or predictive? The expansion of the universe, according to Dr. Faulkner, is definitely explanatory and not predictive, as general relativity suggests either an expanding or contracting universe but predicts neither. So general relativity can explain an expanding universe but it does not necessarily predict it. Concerning the lite elements of the universe, this also appears to be more explanatory rather than predictive. Again, Dr. Faulkner states, "The Big Bang cosmology does predict the abundances of lite elements, but most people fail to realize that information concerning elemental abundances was input in creating the model." Of the three evidences for the Big Bang Theory, the appearance of cosmic background radiation (CBR) is a "clean prediction of the Big Bang Model." CBR is real and has been confirmed many times and, therefore, denying its existence is not an option. Dr. Faulkner points out, however, that though many claim the COBE results "perfectly match predictions," that is simply not true. Theorists have recalculated Big Bang models to produce COBE measurements and this "hardly constitutes a perfect match. Instead, the data have guided the theory rather than the theory predicting the data."

Because of these "evidences" for big bang, some Christian scientists have "latched onto the Big Bang model" and have attempted to use it as "proof" of a creator. They liken the Big Bang to the first verses in the book of Genesis because the Big Bang involves a "singularity" or a beginning, and this implies a creator. Dr. Hugh Ross, progressive creationist and astrophysicist, even offers Scriptures to support the idea of a big bang. Dr. Ross references eleven different verses in the Bible that allude to a "stretching out" of the universe, citing Job 9:8, Psalm 104:2, Isaiah 40:22; 42:5, 44:24; 45:12; 48:13; 51:13, Jeremiah 10:12; 51:15, and Zechariah 12:1. We will look at progressive creationism later, but clearly most naturalistic scientists would argue a big bang event does not offer proof of a creator. There is, however, a problem that Big Bang theorists now acknowledge. That problem is: given the estimated age of the universe at 13.8 billion years and the assumed age of the earth at 4.5 billion years, there simply is not enough time, even in their own theoretical construct, for life to have emerged. For this reason, theories of expanding and contracting universes have now been offered, with an infinite number of universes postulated in order for enough time to allow for life.

Another problem physicists have with the Big Bang model goes back to the very beginning, once again. They have no explanation for what occurred before the Big Bang or during the first 10^{-43} seconds of age. In fact, this period of time (10^{-43}) is the topic of significant speculation amongst astrophysicists. These physicists hypothesize that some time before the universe was 10^{-43} seconds old, different quantum physics had to exist as conditions very early in the beginnings of the universe cannot be explained by current relativity and quantum theory.

The question still becomes: How does something come from nothing and where did space-time come from? Dr. Ross states, "There is no escape from a transcendent creation act" even if one believes in a big bang. There is "no difference from the creation statements of Hebrews 11:3 and Genesis 1:1, where the universe that we humans can detect is said to have been made from that which we humans have no possibility of detecting." Also, if absolute nothingness spontaneously generates space-time, matter and energy, the principle of cause and effect is violated, which further undermines the entire foundation of science, mathematics and logic. Absolute nothingness implies a zero information state, which fails to define when information began; and finally appealing to an infinite number of universes is simply an absurdity, can never be proven or tested, and is "a flagrant abuse of probability theory."

Other Cosmologies

The other cosmologies we will elaborate on all involve a creator or intelligent designer, though most accept an old earth theory. We might ask: Why? There is an attempt to merge creation, especially Biblical creation, with modern day scientific thoughts. Also, "old earth scientists" think the evidence supports a universe of billions of years and look at the book of Genesis from a different point of view.

Broadly speaking, non-naturalistic cosmologies can be categorized into the intelligent design school, old earth creationism, and young earth or biblical creationism. We will look at each of these with some detail, keeping in mind that there is some overlap as some biblical and old earth creationists are in the intelligent design movement as well.

Intelligent Design

The intelligent design movement is very diverse, as is evident from the previous statement. The early roots of intelligent design may be said to have their beginnings with William Paley, author of *The Divine Watchmaker*, who was in the natural theology movement, which sought to support the existence of God through nature. Paley reasoned, if you came across a watch in a field, your assumption would be there had to be a watchmaker. Over the last several decades, the intelligent design movement has come into being partly because of the explosion of knowledge in the fields of biology, chemistry, and physics.

The modern ID movement coalesced based on two events: first, was the publication of Michael Denton's book, *Evolution: A Theory in Crisis* in 1985, which shocked scores of scientist, one being Michael Behe, and the second was the publication of Dr. Phillip Johnson, former U.C. Berkley law professor's book, *Darwin on Trial*. Johnson also known as the "father" of the ID movement, then called a meeting of prospective advocates to join forces in providing a secular response to neo-Darwinism, out of which eventually grew the modern ID movement and its flagship organization, the Discovery Institute.

There are many proponents of intelligent design today, the chief of which might be said to be Stephen Meyer, William Dembski, J. Richards, Paul Nelson, Casey Luskin, Douglas Axe, Jonathan Wells, and Michael Behe. These men and many other scientists involved in ID are highly credentialed and widely published in peer reviewed journals as well as books, with most holding Ph.D.s from secular universities. Some scientists, such as Dean Kenyon and Guillermo Gonzales, have been ostracized by the scientific community for holding their views. This was chronicled in the film, *Expelled: No Intel-*

ligence Required, starring Ben Stein. The Discovery Institute in Seattle is the headquarters of the intelligent design movement.

Before discussing the primary components of ID, it must be emphasized that ID is not necessarily a religious movement and ID scientists come from all types of backgrounds religiously, from Judaism, Islam, Catholicism, and Protestantism. Some ID scientists are even agnostic. As a consequence, the biblical account of creation is not a primary component of ID. Some in the movement do accept the Bible as the inerrant word of God, though most would take a more allegorical approach to the book of Genesis. At least three of the Discovery Institute fellows, Paul Nelson, Nancy Pearcy and John Mark Reynolds, are in fact, young earth biblical creationists.

The basic premises of ID are that the universe is finely tuned and could best be explained by deferring to a designer of very high intelligence; hence the name of the movement. Sir Fred Hoyle, a predecessor of the ID movement, said, "A common sense interpretation of the facts suggests that a super-intellect has monkeyed with the physics as well as chemistry and biology and that there are no blind forces worth speaking about in nature." ID scientists believe that the laws of nature and the physical constants are there for the existence of man. According to Stephen Meyer, chance alone cannot explain the anthropic principle and we are not here just because of sheer luck. The best and, hence, most likely and most scientific explanation of the universe is that it had a "generator," that is God.

At its basic foundation, ID is rooted in information theory. William Dembski's excellent book, *The Design Inference*, masterfully and eloquently written, deals with information and probability theory; specific and highly improbable events require dissemination of information, and information implies intellect.

William Dembski

Interestingly, the late Carl Sagan, an atheist, unintentionally lent credence to the Intelligent Design movement in his fictional book *Contact*. In it scientists receive from SETI a "message" in the form of a series of prime numbers, which was an indication of extra terrestrial intelligence. If this really occurred, you and I would agree with that premise because a series of prime numbers conveys specific non-randomized information and only an intelligence would have sent them. DNA has infinitely more information than a series of prime numbers, yet evolutionists totally ignore the intelligence right in from of them!

Because of the diversity of the ID movement and the attempt to remain neutral, ID scientists have their own views about origins; therefore, the ID movement makes no definitive statements about the origins of the universe, the origins of man, the age of the earth, their view of the Bible, and other biblical concepts such as Noah's flood. Should a Christian then disregard the ID movement? I think not. This would be like throwing the baby out with the bathwater. For example, the ID movement has produced a significant amount of literature and books and multimedia sources that support a biblical creationist viewpoint. Also, as Dr. Georgia Purdom has written, the ID movement "makes clear that Darwinism-naturalism is based on the presupposition that the supernatural does not exist, thus affecting the way one interprets the scientific evidence."

The main problem from a Christian's point of view is its "divorcement of the creator from his creation," which cannot be done with ID. Dr. Purdom states, "All problems inherent with ID stem from this one problem." Theological naturalism failed in the 19th century because of this, resulting in deism, which excluded the Bible completely, attempting to "know god only" through nature and human reason.

Old Earth Creationism

This brings us to old earth creationism, which has four basic schools: gap theory, progressive creationism, framework hypothesis, and theistic evolution. We will discuss progressive creationism mostly and just briefly touch on the other theories. The most popular of the many "gap theories" is that of the "ruin reconstruction hypothesis" first proposed by Presbyterian minister Thomas Chalmers in 1804. Chalmers taught that, in Genesis 1:2, "the earth was without form" should be translated, "became without form and void" because of Satan's fall. God then destroyed the world in "Lucifer's flood" (the first world wide flood) and reconstructed the earth as we know it today. Gap theorists believe that life was formed millions of years ago by God and then wiped out by this "Luciferian flood" between Genesis 1:1 and 1:2 only to be recreated again, as detailed in the rest of Genesis. In this model, the universe was created in the first two verses of Genesis and the earth created secondly; thus, the initial events took millions of years with the rest of creation taking just days.

Framework hypothesists believe man's body developed from ape-like predecessors through God-directed random mutations and natural selection. Man became truly human when God supernaturally placed the image of himself into man. Days of creation then are not literal days but millions of years.

Theistic evolutionists hold that God created the first life in the form of a single-celled organism. This single-celled organism mutated, and through natural selection and millions of years all various life forms came into being, guided by God. Obviously, the "days" of creation are epochs of time in this model. To me, theistic evolution is the biggest copout of all the old earth creation theories and is, of course, neither biblical nor scientific.

Perhaps the most popular old earth creation hypothesis and the one with the most backing of creation scientists is progressive creationism. Its most visible and noticeable proponent is Dr. Hugh Ross, whom we have seen already in this chapter. Dr. Ross is an astrophysicist from the University of Toronto with postdoctoral work conducted at California Institute of Technology. He is the founder and president of "Reasons to Believe," hosts a weekly broadcasted television series, and is the author of several books including *Creation in Time, Beyond the Cosmos, The Creator and The Cosmos*, and *The Genesis Question*. Dr. Ross also has a weekly live nationally broadcast radio program, and is seen on several DVDs including some produced by the John Ankerberg Organization. He is a soft-spoken man, and seems very genuine and sincere, though I certainly disagree with his conclusions.

Progressive creationists believe that God created life in spurts and as some of these became extinct, he made other life forms to take their place. God made man 10,000 to 60,000 years ago to replace soulless hominids, which may have seemed human but were not. In their view, the universe was created by God through the "Big Bang" and the earth was made some 4.5 billion years ago. Progressive creationists believe in the "inerrant" word of God, but assert that the bulk of Genesis involves figures of speech and is not to be taken literally. The flood of Noah was a local event and geologic layers were made over millions of years, according to progressive creationists. Dr. Ross and other progressive creationists try to marry current scientific thought with the book of Genesis. They not only believe science confirms a big bang but, in fact, the Big Bang offers proof of a creator, as we have already seen. There is in fact some logic to this, which we will look at more later, but the basic premise is: If there was a big bang, there had to be a beginning, and a beginning of something implies a creator.

Young Earth Creationists

The next cosmology we will discuss here is, of course, biblical creation. This is the belief of most of the conservative Christian groups in the world, and its leading support organizations are: The Institute for Creation Research, founded by Dr. Henry Morris and headed now by his son, Dr. John Morris; Answers in Genesis, founded by Ken Ham, and editor of the *New Answers* books; The Creation Resource Foundation, founded by Dennis Petersen; and Creation in Symphony founded by Dr. Carl Baugh, president of the Creation Museum in Glenrose, Texas.

Obviously, biblical creationists believe in a young earth and a literal interpretation of the book of Genesis. God created in six, 24-hour days, the heavens and the earth and all life forms in it. The age of the earth is approximately 6,000 to 10,000 years old. Geological layers are primary evidence of a world-wide flood, occurring about 4,500 years ago in the time of Noah and the fossil records are further proof of this event.

One should be aware that young earth creationists do not all hold the same views regarding other biblical doctrines. For example, many, if not all of the above-referenced organizations, are pre-millennialists in nature and have Calvinistic views. In their view, the earth will one day be restored to its original perfection (i.e. a new heavens and a new earth). They seem to ignore obvious Scriptures such as 2 Peter 3:10, which states that the earth and its works will be "burnt up with the coming of the Lord," and 1 Corinthians 15:24, "Then comes the end, when He delivers up the kingdom to the God and Father, when He has abolished all rule and authority and power."

A Few Other Cosmologies

We have just looked at the four basic cosmologies seen in the Western world: 1) naturalism, 2) intelligent design, 3) old earth creationism, and 4) young earth creationism. Now we are going to review three other cosmologies that are not exactly old earth or young earth but rather a permutation encompassing a bit of both. In essence, these cosmologies have a commonality in that they in one form or another attempt to incorporate the evidences of an old earth or old universe with the account found in Genesis. As we have seen, the three primary evidences of a big bang with the obvious interpretation of an old universe are the abundance of lite elements, an expanding universe, and the presence of cosmic background radiation. Also, the apparent distances of faraway galaxies and stars, and the atomic radioisotope dating system, seem to indicate a very old universe. These next three cosmologies incorporate these evidences, yet allow for a creation that occurred in six days as the Book of Genesis states.

Historical Creationism

In the book, *Genesis Unbound*, published in 1996, theologian Dr. John Sailhamer argues that a common, modern understanding of the first two chapters in Genesis is simply wrong. Sailhamer, in his book, approaches Genesis from a textual and biblical argument rather than scientific. He believes the first verse in the book of Genesis, "In the beginning, God created the heavens and the earth" refers to an indefinite period of time. During this period, God created the entire universe as well as the earth that man would inhabit, and the processes by which these were created are simply not stated in the book of Genesis. The second act of God, beginning in Genesis 1:2 and ending in Genesis 2, pertains to a much more limited period of time and scope. In these verses, which detail the first six days of the second act of God, the land is being prepared for man and woman and is the same land promised to Abraham and his descendants. These are six literal days. In particular, chapter 2 of Genesis provides a closer look at God's creation of the first human beings. Sailhamer states, "God creates a people, he puts them in the land he has promised for them, and he calls on them to worship and obey him and receive his blessings."

Sailhamer's approach, according to himself, points to a "proper understanding of the first two chapters of Genesis" but it also "enables us to live in peace with findings of modern science." He is not dogmatic in his belief yet approaches the text from a scholarly point of view, especially looking at the meaning of several key Hebrew words. He affirms that good and godly people can certainly disagree with him.

One of the key words Sailhamer studies is the Hebrew word translated "beginning," *rēʾshît*, which could allow for a long period of time but does not necessitate it. The beginning would allow for many biological eras. When humans are created on the sixth day, "dinosaurs already could have flourished and become extinct." He points out that this same Hebrew word, *rēʾshît* (beginning) is not always used as a starting point but refers to a period of time as seen in Jeremiah 28:1 when the beginning of Zedekiah's reign included events that lasted some four years. There are two specific Hebrew word for "start" and these words are *riʾshônāh* or *tĕchillāh* and are used throughout the book of Genesis but not in verse 1. Further, Sailhamer points out that the phrase *"first day"* in Genesis 1:5 was added by translators and should read, "And there was evening and day, one day." In other words, many days, eons, or epochs could have preceded the one day referred to in Genesis 1:5, with the one day in this case referring to the beginning of God's land preparation for humankind. Evidence for this, from Dr. Sailhamer's point of view, is that human life in the geologic record is quite recent in history compared to the geologic record of other, extinct animals. In verse 2, the phrase *"formless and void,"* according to Sailhamer, could better be interpreted "empty and nothing" or "fallow and indistinct." This would convey the idea of an uninhabitable wilderness, again deferring to the Hebrew interpretation of the phrase *tōhû wābōhû*, which is used in Jeremiah 4, 23 through 26, where it is also interpreted formless and void but clearly refers to a wasteland area. Dr. Sailhamer also makes a distinction between the word *"create"* in verse 1 and the word *"made"* in the following verses. This word *"made"* is used meaning "to appoint" and "to argue." He likens this to the concept of making a bed. So God created the universe in verse 1 and then fixed the sky on the second and third day. Before this, there was a dense fog on the earth which light would eventually transform. So, once again, Day 1 is not the beginning of time but the beginning of the first days of the week in which God prepared the promised land for man and woman. The phrase then *"let there be light"* means, according to Sailhamer, "Let the sun rise." This expression, he points out, is also used in Genesis 44:3, Nehemiah 8:3, and Exodus 10:23.

Genesis 1:6 refers to an expanse in the midst of the waters, with this word *"expanse"* coming from the Hebrew word *rāqia'*, and per Dr. Sailhamer refers to the clouds in the sky and is used similarly in Proverbs 8:28.

For those who question the first verse of Genesis being the creation of the entire universe, Day 4 poses the biggest issue. As you know, in Day 4, Genesis states that God created the lights in the sky, i.e. "luminaries" being the literal word. For Dr. Sailhamer, this is important. In his view, the purpose of the luminaries, expressly stated in verse 14, is to "be for signs and for seasons and for days and years." So the stars were not created on Day 4 but, again, were fixed or set in order. On Day 5, when God created living creatures or beings, these were for the promised land where man was about to be created himself, which obviously took place then on the 6th day.

For those who would argue that Exodus 20:11 proves that God created the entire universe and the world in six days, Dr. Sailhamer points out that the verse does not say God made the heavens and the earth in six days, but rather that God made the heaven and the earth, the sea and all that is in them in six days. The term "heaven and earth" used in Genesis 1:1 refers to the cosmos; the terms "the heaven and the earth and all that is in them," from his point of view, refers to the creation of the Garden of Eden, and hence, the promised land that was given to man at the beginning of creation, and eventually given back through Abraham and his seed.

As one can see, John Sailhamer's "historical creationism" is a rather unique interpretation of the first two chapters of Genesis. It clearly has not received wide acceptance and I certainly don't agree with it, but he at least tries to support his hypothesis through biblical and scholarly research.

White Hole Cosmology

White Hole cosmology was formulated by creationist Dr. Russell Humphreys in the early 1990's. Dr. Humphreys holds a Ph.D. in particle physics from Louisiana State University, is now retired from his research work, and currently works with the Institute for Creation Research. Dr. Humphreys's

A White Hole

White Hole cosmology attempts to give a scientific explanation for the "light years problem" young earth creationists face. The light years problem stems from the fact that stars we see in our sky are judged to be millions and in some cases billions of light years away; yet if the earth and the universe are young, that is, approximately 10,000 years old, how can we be seeing stars that are so far away from us?

Past explanations from young earth creationists for what we will call the time-light-travel issue were simply unsatisfactory from Dr. Humphreys's point of view. One of the explanations given in the past was the inaccuracy of the actual measurements of the stars. This was really a poor explanation, and both Dr. Humphreys and Dr. Danny Faulkner, an astrophysicist, say that such a hypothesis ignores how accurate these measurements really are and is simply not an option.

The most common approach to the time-light-travel enigma appeals to the concept of a mature creation. According to it, young earth creationists believe that God created the universe and the earth as well as the animals and man himself in a mature state. In other words, Adam was created full-grown and fully functioning, as were trees, other animals, and the universe itself. As we have seen, the stars were created on the 4th day before mankind was created. For them to serve the purpose of defining signs and marking seasons when Adam was created on Day 6, he would clearly need to be able to see them. The argument then goes that perhaps light was created in transit on its way toward the earth and visible by at least the 6th day; therefore, the universe has only the appearance of age. Since relativity theory states that nothing travels faster than light, this would therefore be the logical explanation for this phenomenon (interestingly Big Bang theory does allow for "space to travel faster than light"). This reasoning may have some appeal concerning the earth, but for many this explanation for the old universe is less than satisfactory. Danny Faulkner has explained, "If the Universe appears young, then it should not appear old, and if it appears old, then we should not expect the Universe to appear young. We cannot have it both ways."

With the mature creation hypothesis, the stars never emitted light that we are seeing now but instead the light itself was created in transit and only appears to have been emitted. Dr. Faulkner points out two problems with this. First, the creation of light in transit questions the reality of stars; they are illusionary. The bigger problem though is these astronomical bodies contain very detailed information and through the process of spectronomy, we can determine their temperatures, composition, motion, and many other characteristics. If light were only created in transit, why would it have so much information? This is a big dilemma for the mature universe hypothesis.

Clearly, the mature universe hypothesis makes no predictions and therefore cannot be considered scientific. This certainly does not mean it is necessarily wrong, but makes it untestable and hence a philosophical argument, not scientific.

Another young universe explanation for the time-light-travel problem is the so-called Setterfield Hypothesis. Australian Barry Setterfield proposes that light traveled at a much faster speed in the past than it does now, maybe even with infinite speed. This is a fanciful idea but ignores the constancy of light and the permanency and permeability of free space. Mr. Setterfield's argument is that in the past the speed of light has been measured differently. The problem is not in the speed of light, however, but in man's measurements and instrumentation. Since 1960, the speed of light has never varied with modern instrumentation. As a result, Setterfield's explanation leaves much to be desired when attempting to answer the light-time-travel issue, and this is especially true from Dr. Russell Humphreys' point of view.

Dr. Humphreys, a creationist, began looking at a more viable explanation, culminating in the publication in 1994 of his book, *Starlight and Time*. From this book comes the concept of the White Hole hypothesis.

The white hole hypothesis has two important starting points: 1) Relativity is accurate and real, and 2) The cosmologic principle, also referred to as the Copernican Principle, is flawed and there can be a center of the universe.

As to the first point, the reality of relativity is that time is not absolute. For example, clocks run faster at higher altitudes and this has been proven with atomic clocks. The clock at the Royal Observatory in Greenwich, England ticks 5 microseconds slower than the clock at the National Bureau of Standards in Boulder, Colorado. Which time is correct? The answer is they both are; it is their differences in altitude, and hence their frame of reference differences, that explain the phenomenon of gravitational time dilation. Dr. Humphreys states, "The effect applies to the rates of all physical processes" including aging and decay. Dr. Humphreys asks, "When time is measured during the creation week, which frame of reference is used?" This seems to me to be a valid question. Using a white hole cosmology and the two starting points we've just discussed, Dr. Humphreys has developed a theory that states light has ample time to travel in six days when using "the earth's frame of reference (Earth Standard Time)." So for "the math to work" the earth must therefore be the center of the universe, which violates the cosmologic or Copernican Principle, which is really just an arbitrary assumption to begin with.

Dr. Humphreys deduces that the visible universe was once inside the event horizon of a white hole, and if the universe is bounded then sometime in the past the universe must have expanded out of a white hole. That white hole is now gone (as are all white holes theorized by physicists). This would have all occurred from Dr. Humphreys's point of view before the end of the 6th day.

To understand the white hole cosmology, let's look at what a black hole is, since, in essence, a white hole is its opposite. Black holes are areas of the universe where gravity is extremely strong, so strong that even light itself cannot escape its gravitational pull. You cannot see black holes, after all they are black, but there is a significant amount of evidence that they are real, and the center of our own galaxy, the Milky Way, may contain a black hole. At the edge of a black hole is its event horizon, where anything approaching it can no longer escape.

White holes are purely theoretical and according to physicists no longer exist, but their effects would have been quite different from the black holes and most physicists believe they did in fact exist at one time. White holes would release matter outwards by an anti-gravitational effect and would have appeared extremely bright as a result. Also, as you would expect that the horizon event of the white hole would not allow anything to return to it, all matter would therefore accelerate away from the white hole. Time would be viewed very differently from the edge of the white hole or its event horizon and this is the basis of Dr. Humphreys' cosmology.

The main point in this theory is that according to general relativity, time effectively stands still at the event horizon. "If you were standing on the Earth as the event horizon arrived, distant objects in the universe could age billions of years in a single day your time," says Dr. Humphreys.

The White Hole cosmology has a certain appeal and retains the literal six days of creation we read about in Genesis, though a few details clearly remain to be worked out. Both the White Hole and Big Bang Cosmologies make assumptions. The former assumes a bound universe with the earth as the center and the latter the exact opposite. Dr. Humphreys's cosmology, however, has the great advantage of being much more biblical!

Genesis and the Big Bang
The final cosmology we will examine comes from Dr. Gerald Schroeder, and from my point of view incorporates a bit of the white hole hypothesis and the historical creationism of Sailhamer, yet is quite unique. Dr. Schroeder comes from a Jewish background and is an applied physicist and theologian, and received undergraduate and doctoral degrees from the Massachusetts Institute of Technology. His research has been published in numerous scholarly journals. His cosmology is detailed in his book, *Genesis and the Big Bang*, published in 1990. In it, he describes how, in his opinion and based on a scholarly approach, the book of Genesis and modern science as it pertains to cosmology are in complete harmony. Dr. Schroeder states, "The biblical narrative and the scientific account of our Genesis are two mutually compatible descriptions of the same identical reality."

The essence of Schroeder's hypothesis goes back to the rate of time's passage and how objects that were felt to be fixed in a point of time in the past now are in actuality dependent on the relation of the observer and the observed. Dr. Schroeder views the earth and the universe as old from man's point of reference, but very young from God's. In other words, from "the beginning" the appearance of man did take six days and 15 billion years "simultaneously" starting at the same instant and finishing at the same time.

How can this be? Here is the key to Schroeder's cosmology: "Until Adam appeared, on the 6th day, God alone was watching the clock." I must emphasize this is not the gap theory we looked at briefly before. Though a bit difficult to understand, because of the concept of relativistic time, the universe contains billions of cosmic clocks, each ticking at its own locally current time. According to Schroeder, it was at the moment of Adam that the time-space relationship of God and man would be the same; therefore, the genealogies of Genesis would be accurate.

Once again, Dr. Schroeder states, "When the Bible describes the day-by-day development of our universe in the six days following the creation, it is truly referring to six 24-hour days. But the reference frame by which those days were measured was one which contained the total universe."

From Dr. Schroeder's point of view, the first verse of the Bible incorporates a singularity, the Big Bang, which was the beginning of time and space and matter. Contrary to naturalism, Dr. Schroeder believes something could come from nothing only because of a creator or God who spoke it into existence. From his point of view, cosmic background radiation, the expansion of the universe, and the abundance of light elements stem from this event, spoken into existence by God. Furthermore, all the constants seen in the universe, even the very size of the universe, are determined by God and for the existence of man.

In Schroeder's cosmology, the evening and morning phraseology of Genesis is better characterized as disorder (evening) and order (morning) with order in this sense representing highly ordered structures transcending the second law of thermodynamics (entropy). The term, *"It was good"* in Genesis refers to this order.

Though Dr. Schroeder may not be considered technically a theistic evolutionist, in that he understands organisms even as simple as a bacterium could not have occurred randomly even if the age of the universe is 15 billion years, he nonetheless believes that God could essentially direct evolution.

As for Day 4 of creation, also a problem in Dr. Schroeder's cosmology, he postulates that the light created in the first day was translucent but not transparent through Earth's cloudy, vaporous atmosphere. As a result, individual luminaries would not be visible on Earth. That these events are being described from an earthly point of view is "made clear by the reference to the moon as a great luminary (Gen. 1:16)." Only from the earth could the moon have been seen as a luminary, because from outer space it would be too close to the earth to be viewed in this way.

Again, though not strictly a theistic evolutionist, Dr. Schroeder does believe in a common ancestry of man with animals, citing the most compelling evidence being the similarity of the genetic material found among all forms of life. He admits such "evolution" would need to be punctuated, but as we have seen in our previous study, similarity does not require common ancestry but rather implies a common maker or designer; and there is simply no evidence of punctuated evolution.

For Dr. Schroeder, some 5,700 years ago a "quantum change" occurred and man was given an additional spirit or soul, the *něshāmāh* (*breath* of life, Gen. 2:7). At this point, reasoning, speech, and other capabilities of mankind were imparted to him by God. This of course is a very similar concept to what we have seen from the progressive creationists.

How would I characterize Dr. Schroeder's hypothesis? For me, there is some merit to the concept of relativistic time and measuring time from one's point of reference. God, however, transcends time and space; therefore, how and where would be His point of reference?

Final Thoughts

Dr. Robert Eckel from the book, *In Six Days*, has stated, "Creation in six days is not an intellectual argument that can be won by in-depth and repetitive examinations of the scientific evidence available." An unbelieving world will never give credence to the things those more spiritual would believe, and possibly was not even intended to do so. 1 Corinthians 1:18 through 29 is a good dissertation on this very subject. "For the word of the cross is to those who are perishing foolishness, but to those who are being saved it is the power of God." Verse 19, "For it is written, I will destroy the wisdom of the wise and the cleverness of the clever I will set aside." Verse 22 states, "For indeed Jews ask for signs and the Greeks search for wisdom but we preach Christ crucified, to the Jews a stumbling block and to the Gentiles foolishness." Having said this, however, the position of the apologists needs to be scientific and sufficiently well-developed academically in order to interchange information and be implemented in the world's terms, understanding that proof of either position can never be wholly substantiated.

Hebrews 11:3 states, "By faith we understand that the universe was formed by God's command." I believe in the cosmology described in the book of Genesis. But the universe does appear to be old. We have studied the evidences including the light-time-travel phenomenon, cosmic background radiation, an expanding universe, and the abundance of light elements. One could choose to ignore the science and that would certainly be okay but it is not for me.

Could God have created the universe in six days? Could God have created it in 14 billion years? Could he have created it in 10^{-43} seconds? The answer to all three of the preceding questions is yes, of course. So the question is not what God could have done but what actually occurred; and it may be and probably is, that mankind is just incapable of comprehending how it all was done.

The scientific evidence is that God used "processes" (such as gravity, light, etc.) that He spoke into existence. God has set man apart from animals and I believe that at least a portion of Genesis 1:26 entails the part of man that attempts to discern these processes as God did not reveal them to man.

The Bible, in particular the book of Genesis, was written, probably by Moses, through the guidance of the Holy Spirit (2 Tim. 3:16) and was written for all ages. Early man would be able to comprehend it in one way, and modern man, having discovered in part some of these processes, would understand it a bit differently. From my point of view, understanding the light-time-travel enigma as well as the other phenomena including an older universe, lies within the domain of relativistic time which has been experimentally verified many times. Dr. Russell Humphreys's white hole cosmology makes some intuitive sense to me, though his theory is certainly not fully developed. If the white hole cosmology is true, then stars and galaxies are billions of light years away, thus giving credence to an old universe, yet were still created in six days no matter how contradictory that may seem.

We will come back to this discussion again when we study evidences of a young earth, but suffice it to say for now there are evidences for a young and old earth just as there are for the universe. Regardless of whether the earth is 4.5 million years old or 10,000 years old, there is still simply not enough time for evolution to have occurred from naturalistic processes. The key it would seem is in differentiating between data and interpretation.

Dr. Elaine Kennedy, a young earth scientist with a Ph.D. in Geology, has pointed out the data in the geologic record can be interpreted to support either a long (millions of years) or a short (a few thousand years) history on Earth. Some data are better for one, some better for the other. How one interprets the data is based upon his or her perspective and belief, or not, in the Holy Scriptures.

I would like to make one further comment, and that is God does not require that we understand all the nuances of his creation in order to be a Christian and for one to require such is a burden that is unnecessary and counterproductive. The requirement for Christianity is a belief in the deity of Jesus Christ and adherence to his teachings.

Chapter 17

A Fine-Tuned Universe

Ever since the 16th century, Polish astronomer Nicholas Copernicus shattered the notion of an earth-centered universe, our planet has been relegated to a position of mediocrity. What has been described as "the principle of mediocrity," also known as "The Copernican Principle" and "The Cosmologic Principle," basically says that our earth is nothing special. Before Copernicus, mankind had always believed that the earth was the center of the universe and that the sun and other planets revolve around it. The Copernican Principle makes several predictions. The first prediction is that Earth, while it has many life-permitting properties, isn't exceptionally suited for life in our solar system and other planets in the solar system probably harbor life as well. The Copernican Principle also predicts that our sun is fairly ordinary and a typical star; and our solar system as well is very typical and we should expect many other solar systems to mirror our own. Further, one would expect a lot of planetary configurations that are consistent with the presence of biological life. The Copernican Principle also predicts that variables like the number and types of planets and moons have little to do with the existence of life in a planetary system. Further, our solar system's location in the Milky Way is unimportant and our galaxy is not particularly exceptional either. Finally, the Copernican Principle would predict the universe is infinite in space and matter and time, and the laws of physics are not specially arranged for the existence of complex or intelligent life.

As a result of this Principle of Mediocrity, some in the past have predicted that our own galaxy contains many advanced civilizations. Frank Drake and Carl Sagan formulated the Drake Equation based upon several educated guesses and came up with the startling conclusion that intelligent life was common and wide-spread throughout the galaxy, and indeed Sagan himself estimated the number of planets with intelligent life as in the millions in our galaxy. As a result, by the end of the 20th century, interest in extra-terrestrial life permeated our society, both in fictionalized creations in movies such as *2001: A Space Odyssey, Close Encounters of the Third Kind,* and *Contact,* as well as serious scientific search for extraterrestrial life through a series of large radio wave receivers in a project known as SETI (Search for Extraterrestrial Intelligence). Early on, Frank Drake himself transmitted radio signals into space hoping that an alien civilization might intercept them. Now researchers spend their time trying to detect intentional or unintentional radio transmissions from extraterrestrials. Although early on there were some false readings, which later turned out to be pulsars from distant stars, to date no intelligent life has been identified outside our planet. Though certainly it is possible at some point mankind may indeed discover extraterrestrial life, it remains a bit of a puzzle that we have not found evidence to date when supposedly millions of planets in our galaxy contain it. In 1950, Nobel Laureate and physicist, Enrico Fermi, argued that if there are extraterrestrials, we should have heard from them by now. In what has been called Fermi's Paradox, which is no paradox at all unless one assumes extraterrestrial life exists, the argument is that if there are numerous other intelligent civilizations in our Milky Way, some of them would surely have had a head start on us and eventually run out of room or encountered some reason, be it a hazard or simply curiosity, which would have encouraged their migration; and we should have heard from them by this point.

Some well-respected scientists are now questioning the Copernican Principle altogether. Fifteen years ago, Peter Ward and Donald Brownlee's seminal book, *Rare Earth*, argued there was strong evidence that our earth seemed perfectly placed in the universe to support life, and seemed unique among other discovered planets in this capacity. Brownlee, an astronomist at the University of Seattle, and Ward, a geologist and Curator of Paleontology at the University of Washington at Seattle, give strong evidence in their book why complex life would be very uncommon in the universe. Planets may be common, but it is difficult to find planets like our earth that could support life as we know it. They state in the preface of their book, "Earth seems to be quite a gem – a rocky planet where not only can liquid water exist for long periods of time," thanks to Earth's distance from the sun, as well as its possession of a tectonic "thermostat that regulates its temperature," but where water can be found as a healthy ocean – not too little and not too much. Our planet seems to reside in a benign region of the galaxy where comet and asteroid bombardment is tolerable and habitable-zone planets can commonly grow to Earth size. "Such real estate in our galaxy – perhaps in any galaxy – is primed for life. And rare as well." Guillermo Gonzalez (an astronomist) and Jay Richards (a philosopher and theologian) make similar arguments in their book, *The Privileged Planet*, published in 2004. Gonzalez and Richards are especially amazed by the fact that our earth is not only rare, but also situated in our galaxy that allows for discovery. In their introduction to the book, they state, "Our claim is that Earth's conditions allow for a stunning diversity of measurements, from cosmology and galactic astronomy to stellar astrophysics and geophysics; they allow for the rich diversity of measurement much more so than if Earth were ideally suited for, say, just one of these sorts of measurement." In other words, it seems as if man was placed on Earth with the ability to discover his place in the universe. Interestingly, although Ward, Brownlee, Gonzalez, and Richards all understand the uniqueness of the earth, the significance of this is viewed quite differently by them. Ward and Brownlee are naturalists and unwilling to accept intelligent design as an explanation but rather defer to the Anthropic Principle, which we will discuss in more detail at the end of this chapter. Gonzalez and Richards, on the other hand, see design in the cosmos evidenced by the correlation of meaningful pattern throughout it. They state, "Design provides just such a tidy explanation here. Think of it this way: If the physical universe were designed so that any observers would find themselves in an environment conducive to many diverse scientific discoveries, then the correlation would be just what they would have expected. Although this evidence might not prove the cosmos was designed it would surely confirm it." They "assume" that the universe is designed at least in part to allow intelligent observers to make discoveries; the correlation between life and discovery we observe is what we expect.

Rare Earth
So what is so rare about our earth? Brownlee and Ward summarize very nicely the factors that make our earth so rare indeed. First, Earth is situated in a galaxy in the universe that allows life to exist. Where life would unlikely occur would be in what Brownlee and Ward refer to as dead zones. These dead zones include distant galaxies which would not have enough metals for formation of Earth-size planets, globular cluster galaxies that are too metal poor to have inner planets as large as Earth, elliptical galaxies whose stars are also too metal poor, and small galaxies as you might guess also too metal poor. The centers of galaxies would also be dead zones, because energetic processes would impede complex life. The edges of galaxies would contain many stars that are, again, too metal poor. Planetary systems with "hot Jupiters" would drive the inner planets into the central star, making it uninhabitable. Planetary systems with gigantic planets in eccentric orbits

would produce environments too unstable for higher life, and future stars, should they occur, would not have enough uranium, potassium, and thorium to provide sufficient heat to drive plate tectonics.

On the other hand, our earth possesses rare factors that allow for life to exist and continue to exist. We seem to have the right type of star which produces not too much ultraviolet light; we seem to be the right distance from the star to allow for liquid water near the surface, but far enough to avoid tidal lock. Our earth is the right size and hence has a very stable orbit. It also retains atmosphere and oceans and enough heat for plate tectonics, a concept we will shortly detail. Jupiter, our distant neighbor, is important in clearing out comets and asteroids. It appears to be not too close, not too far. Mars, a nearer neighbor, seems small enough, inhabitable enough to allow for future colonies. The plate tectonics of the earth produce CO_2, built up land masses, enhanced biotic diversity, enable a magnetic field, and provide a thermostat for the earth. Our oceans seem just the right size; not too much nor too little. Our relatively large moon appears to be just the right distance from the earth and helps to stabilize the tilt of the earth, which enables seasons not to be so severe. The earth has just the right amount of carbon for life but not enough for runaway greenhouses. We seem to have very few giant impacts at this point which, of course, would destabilize our environment. The atmospheric properties of the earth maintain adequate temperatures for plant life and animal life. We seem to have the right kind of galaxy and the right position in the galaxy; not in the center or at the extreme edge. Finally, our earth allows for complex plants and animals to exist with just enough photosynthesis to produce oxygen.

Hugh Ross in his book, *The Creator and the Cosmos*, lists 35 factors or parameters for the universe that must be what they are within a finely narrow range for physical life of any conceivable kind to exist. These factors include the strong nuclear force, the weak nuclear force, the gravitational constant, the electromagnetic force constant, the ratio of electron to proton mass, the ratio of numbers of protons to electrons, the expansion rate of the universe, the entropy level of the universe, baryon or nucleon density of the universe, the velocity of light, the age of the universe, the initial uniformity of radiation, fine structured constants, average distance between galaxies, average distance between stars, the decay rate of protons, carbon to oxygen energy level ratio, ground state energy level of helium, the polarity of the water molecule, the size of relativistic dilation factor, and the uncertainty magnitude in the Heisenberg Uncertainty Principle, to name some of these parameters. He further lists 66 parameters for the fine tuning of the Galaxy-Sun-Earth-Moon system for life support. He states that these parameters "of a planet, its moon, its star, and its galaxy must have values falling within narrowly defined ranges for life of any kind to exist." In looking at all these factors, Dr. Ross estimates the probability of obtaining the 128 necessary parameters for life support to be 10^{-166}. This is an extraordinarily high number as the maximum possible number of planets in our universe is estimated to be only 10^{22} power. This would be less than one chance in 10 to the 144 (trillion, trillion, trillion, trillion, trillion, trillion, trillion, trillion, trillion, trillion, trillion, trillion) for even one such planet to occur anywhere in the universe.

Brownlee and Ward have devised what they call the rare Earth equation to determine the probability of life somewhere else in the universe. Looking at the stars and the Milky Way Galaxy times a fraction of stars with planets, times a fraction of metal rich planets, times the planets in a star's habitable zones, times the stars in a galactic habitable zone, times the fraction of habitable planets where life does arise, times the fraction of planets where life or complex metazoans arise, times the percentage of a lifetime of a planet that is marked by the presence of complex metazoans, times the fraction of planets with a large moon, times the fraction of solar systems with Jupiter-sized planets, times the fraction of planets with a critically low number of mass extinction events, equals a probability that approaches zero. To say these numbers seem to contradict the Copernican Principle is a great understatement.

We cannot look at all the various parameters and factors that make our planet unique but I would like to look at the importance of our sun, our moon, plate tectonics, and the planet Jupiter in providing an environment on Earth that is conducive for life.

The Sun

Contrary to the Copernican Principle, our star, the sun, is not just a common star but has properties making it unique to our planet, therefore causing the planet to be habitable. The sun is called a G2 Star and has significant mass putting it in the top 10% of all stars. It contains 70% hydrogen, 28% helium, and 2% of other metals. In essence, the sun, like all stars, is a gigantic nuclear bomb. It converts hydrogen to helium in its core from the pressure of gravity through a process known as nuclear fusion, with the core temperature being 15.6 million K (Kelvin). The temperature at the photosphere, the perimeter of the sun, is much cooler yet obviously still

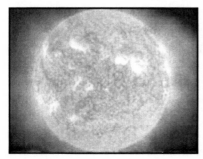

exceedingly hot at 5,800 K. Through a telescope, sun spots can be seen which are cooler areas in the photosphere. Also visible on the surface of the sun are solar flares which produce solar winds, which send out low density streams of charged particles that can be detected on the earth and seen in the polar regions of the earth in the form of the aurora borealis.

The benefits of the sun are somewhat obvious and certainly life could not exist on Earth without it. It provides sources of chemical elements and a steady supply of energy, which in turn allows for photosynthesis which supplies a never-ending source of food for the planet. The earth is somewhat fortuitous in that it rests in the circumstellar habitable zone of the sun, which basically means we are not too close nor too far away for complex life to exist. We are also lucky in that our sun is not too big or too small. Stars too big aren't compatible with life because they change too quickly, resulting in dramatic climate changes. Low mass stars, or M dwarf stars, are not particularly conducive for life either, as planets would need to be in close proximity to these stars, and this would result in severe tides making complex life difficult. Also, water would freeze and solar flares would put out significant radiation, anywhere from 100 to 1000 times more than our sun. Dual star systems don't work and neither do globular star systems for basically the same reasons. Our sun is also highly stable, with light output varying by only 0.1% over a full sun spot cycle that is 11 years. Furthermore, our sun's luminosity or brightness seems especially optimal for extracting information from its spectrum. Finally, the size of our earth in comparison to the size of the sun is optimal as well. If our earth was too small we would have to be closer to the sun to receive enough energy. But too close would produce "tidal lock," where the tides of the planet would always focus towards the sun, which would make the planet uninhabitable.

Guillermo Gonzalez and Jay Richards state in their book, *The Privileged Planet*, the sun also serves as a handy astrophysics lab. Because of the earth's proximity to the sun, we can obviously study it much easier than we can other stars. The sun produces neutrinos, which are nearly massless neutral particles that hardly interact with matter, but through their study scientists have learned a lot about the very small and the very large, while it also offers direct testing of other solar models. Summing it up, Gonzalez and Richards state, "So the sun's local environment seems to offer the best type of habitat for complex life. At the same time, its particular properties disclose vital scientific information more abundantly than many more common types of stars, while also providing us with an excellent example of stars in general."

The Moon

Again, contrary to the Copernican Principle, our moon also seems quite unusual and allows for the existence of complex life. Peter Ward and Donald Brownlee in the book, *Rare Earth*, even go so far as to say, "The

presence of our huge moon seems to be one of the most important and yet most perplexing. Without the large moon, Earth would have had a very unstable atmosphere and it seems most unlikely that life could have progressed as successfully as it has."

Why is the moon so important? Because of its relatively large size, being approximately one-third the size of the earth, the moon through

its gravitational effect exerts a large influence on the earth and its axis. As you may know, the earth tilts between 21.5 and 24.5 degrees on its axis. This tilting of the earth turns out to be extremely important and conducive for a habitable planet. The axis causes lunar tides, it stabilizes the earth's spin, and it slows the rotation rate of the earth. Because of these factors, we have seasonal change. Without it, the surface of the earth would be either extremely cold or extremely hot, making complex life unlikely. The poles of our earth would be in total darkness for half a year and in constant illumination for the other half. And the degree of tilt seen in the earth seems to be perfect as well. Too much tilt would be just as disastrous as no tilt at all.

The moon also is responsible for the tides we experience on Earth due to its gravitational effects with tides also being affected by the sun. Tides drive currents, which help regulate climate by circulating enormous amounts of heat, and without lunar tides our world would look quite different. As Gonzalez and Richards point out, "Seattle would look more like southern Siberia than the lush, temperate Emerald City." As you might guess, the tides on our earth appear to be just right as well because of the size of both the sun and the moon producing sufficient variation but not enough to be so drastic as to inhibit life.

Another benefit of our moon is in its protective ability against excessive bombardment from comets, meteors, and asteroids. Also the size of the moon and its distance from Earth and from the sun allows for inhabitants of Earth to witness periodic eclipses. That in itself may not seem a significant benefit; however, because of these total eclipses mankind has been able to make many discoveries and verify other theories such as discovering the nature of stars, measuring the slowdown of Earth's rotation, and proving Einstein's general relativity theory.

Jupiter

As it turns out, Jupiter also plays a significant role in the habitability of our planet. Jupiter is a massive planet, being some 318 times the mass

of the earth. Visible with even a small backyard telescope, Jupiter has a very exotic exterior, exhibiting some parallel equatorial bands of a whitish color quite distinct from the color of the earth when seen from outer space. Jupiter has four moons, first discovered by Galileo in 1612. Jupiter has no surface but instead is a giant gas ball which gets hotter and denser with depth and is made up mostly of hydrogen and helium in its deep interior, and at its core with millions of atmospheres of pressure hydrogen is actually in a metallic state.

These facts are interesting; but how does Jupiter benefit our planet? By far the most beneficial aspect of Jupiter's existence to Earth is its ability to "clean out" stray bodies such as asteroids through its enormous gravitational effect. Once again, Earth seems to be just the right distance from Jupiter for this effect to be of benefit. If Earth had been a bit closer to Jupiter or Jupiter had had somewhat of a larger mass, this beneficial effect would not have occurred, and as a result the earth would have been bombarded with these astronomical bodies rendering it much smaller and incompatible with life. It also helps that Jupiter has a very stable orbit and this quality is probably rare. Of course, objects still impact the earth today but fortunately for us their impact rarely has much consequence, as Jupiter has seemed to have protected us from the much larger bodies.

Planets the size of Jupiter have been discovered in extrasolar planets;

however, in every case these planets are either too close to their star (hot Jupiters), or their elliptical orbits are too far from the star. In both cases, hot Jupiters or Jupiters too far from their respective star would not be beneficial to Earth-sized planets. To date, no "Jupiters" have been found amongst the hundred or so Jupiter-sized planets mankind has discovered, again substantiating the uniqueness of our Jupiter.

Plate Tectonics

When observing other planets, two things stand out on Earth to make it unique: one being the abundance of water on Earth, the other being the presence of linear mountain ranges, neither of which have been observed on other planets. Although it might not be obvious, without these two intertwined phenomena, life would be difficult on Earth. Plate tectonics play an extremely important role in the habitability of Earth by providing a sort of homeostasis between the earth's oceans and land masses. Plate tectonics are the dominant features that cause changes in sea level, which as it turns out are vital for the formation of minerals that keep the level of global carbon dioxide in check; and, of course, allow for the creation of continents or land masses for animal life to dwell on. Plate tectonics also make possible the production of our magnetic field on Earth, which in turn protects us from potentially lethal influxes of cosmic radiation and solar winds which would eat away at our atmosphere.

Continental drift, the precursor to "plate tectonics" was first scientifically proposed by a creationist, Antonio Snyder, based on the Genesis account of "God gathered the waters into one place" suggesting that there was one giant land mass. He then made and published two maps showing like a jigsaw puzzle what this continent originally looked like. Unfortunately his work was published in 1859, the same year as Darwin's book, thus hurting his publicity. Snyder was ridiculed and rejected (although he was right), nevertheless he was all but forgotten from the geologic sciences, in part due to his creationist beliefs.

So what are plate tectonics? In 1910, American geologist Frank B. Taylor proposed that the drifting of continents caused the great mountain chains on Earth. Mountains are either volcanic or non-volcanic, but before Taylor's proposition how the mountain ranges were formed was a mystery. The idea that continents could "float" over a planet's stony surface was heretofore an unknown concept. Since the time of Taylor, man has discovered that the uppermost layers of Earth, known as the crust or mantle, through the process of thermal convection do in essence float. Scottish geologist Arthur Holmes proposed that the upper mantle acted much like boiling water, producing large "cells of material," this hot material then heating and rising and cooling again eventually to begin a flow parallel to the planet's surface. When the material rises, the convection cells rupture the rigid crust then carry it along regions where the mantle moves parallel to the surface. German meteorologist Alfred Wegener, from his studies of paleomagnetics which allowed the reconstruction of ancient continental positions, revealed the presence of enormous underwater volcanic centers, "Areas where the seafloor literally pulls away from itself."

Continents are composed of granite and basalt, primarily, but because granite is less dense than basalt, the granite-rich continents essentially then float on a thin bed of basalt. To conceptualize this, think of an onion with its various layers and skins, with the continents of granite acting like smudges of different material embedded in the onion skin. The interior of the earth is very radioactive and produces significant quantities of heat. As the heat then rises to the surface this creates liquid rock, and like boiling water this viscous upper mantle rises and moves parallel to the surface for many distances, then when cooled settles down into the depths of the ocean. These gigantic cells carry this thin, brittle outer layer, and this is known as plates, hence our tectonic plates.

The tectonic plates then lie on a bed of decreased viscosity. This allows for the rigid crust to slip over the zone of lower viscosity. Continents then begin to drift. Subduction zones are long straight regions driven down in Earth and when colliding together produce mountain ranges; thus these ranges are a result of collision of plates buckling and crumpling.

Volcanoes occur along subduction zones. Because magma is of lower density than basalt, new rocks are formed by volcanoes which in turn produce mountains. Plates interact with each other by spreading at the center, by grinding along each other, and colliding with each other, which in turn produces earthquakes and volcanoes.

So what's the importance of plate tectonics? First, plate tectonics are felt to be responsible for the production of continents. It is believed that at one time there was only one massive continent on the surface of the earth. Through the process of plate tectonics, probably influenced by the Great Flood of the Bible, this land mass began to separate and "drift" (or sprint if you are a young earth creationist), resulting in the continents that we see today. Scientific evidence has shown continental drift to be a fact and at least partial proof can be seen by looking at the western border of Africa and the eastern border of South America. These borders seem to "fit" giving at least some credence to the concept of Pangea, the term used for this supercontinent.

Another benefit from plate tectonics is that it allows for a divergence of habitats, producing environmental complexity and allowing for the adaptation of the various kinds. This divergence in kinds is certainly beneficial for the earth in that it allows for relative abundance of certain kinds and this biodiversity I believe is a good thing.

Plate tectonics are also responsible for volcanism and if these plates cease to move, volcanism and, for that matter, earthquakes would stop as well. On the surface, this may seem like a good thing; but in fact it is not. There is a dynamic associated with volcanism and plate tectonics, in that plate tectonics can produce volcanism but volcanism in turn can produce plate tectonics. If plate tectonics stop, the earth would eventually lose its continents due to erosion at least based on an "old earth" view. In a "young earth" model with different erosion rates this would not necessarily be true.

In addition to this, carbon dioxide which is removed from the atmosphere by weathering, which is further produced by tectonics, would cease to be removed resulting in a planet freeze. This thermoregulatory quality of plate tectonics is very important in helping to maintain the important greenhouse gases of H_2O, O_3 (ozone), CO_2, and methane. According to Brownlee and Ward, "Greenhouse gas composition and thus planetary temperature are byproducts of complex interactions among a planet's interior, surface, and atmosphere."

Plate tectonics are also responsible for recycling minerals and chemicals locked up in the planet's sedimentary rocks. The CO_2-rock weathering cycle is key to maintaining the earth's temperature and is not known to occur anywhere else in the universe. A significant portion of carbon dioxide (CO_2) is produced into the atmosphere by volcanoes and taken out through limestone formation, thus creating a delicate balance.

Perhaps the most important role of plate tectonics is in producing our magnetic field. Our magnetic field is produced by an inner core of matter (iron) spinning through Earth's rotation creating convection movement and hence magnetism. It is plate tectonics that cause loss of heat that produces these convection cells necessary for magnetism. Why is magnetism important? Our magnetic field reduces "sputtering" of the atmosphere but more important it protects us from serious cosmic radiation that is reflected from Earth because of this magnetism. We are not completely sure why we have tectonics nor exactly how they function but it is important that water be abundant on Earth for tectonics to occur and Guillermo Gonzalez has pointed out that the earth seems to have always had an abundance of water. Interestingly, this coincides with what the Bible says in the first chapter of Genesis. Water is needed for subduction and without it there would be no plate tectonics.

The Anthropic Principle

There is little debate among scientists about the fine tuning of the universe. We have seen that the universe and indeed the earth and our solar system do seem to be unique and finely tuned for the existence of complex life here on Earth. So virtually all scientists, be they believers or not, understand how fine-tuned the universe is. We have noted, Peter Ward and Donald Brownlee have calculated that the odds of complex life on another planet in the universe approach zero. So if most scientists don't believe in an intelligent designer, just what is their explanation? Their answer simply is the Anthropic Principle.

The phrase "Anthropic Principle" was coined by Brandon Carter in 1973, interestingly, during a symposium honoring the 500th birthday of Copernicus. The Anthropic Principle actually has two variants; the strong Anthropic Principle and the weak one. The strong Anthropic Principle is an attempt to explain why the universe has the age and the fundamental physical constants necessary to accommodate conscious life. For its proponents, it is unremarkable that the universe's fundamental constants happen to fall within the narrow range thought to be compatible with life. John Barrow and Frank Tipler believe this is the case because the universe is compelled in some sense, to eventually have conscious and sapient life emerge in it.

The weak Anthropic Principle argues that only in a universe capable of eventually supporting life will there be beings capable of observing and reflecting on it, with a universe not so fine-tuned simply going unbeheld. This type of argument falls back on the multiverse notion, which basically states that given enough universes sapient life would eventually occur. Of course, there is no shred of evidence for a multiverse reality.

Roger Penrose has explained the weak Anthropic Principle very succinctly, "The argument can be used to explain why the conditions happened to be just right for the existence of (intelligent) life on Earth at the present time. For if they were not just right, then we should not find ourselves to be here now, but somewhere else, at some other appropriate time." An analogy to this is our own very existence. The likelihood of any one of us ever being born is very small yet we are here and if we weren't we wouldn't know it anyway.

As you can imagine, the Anthropic Principle cannot be falsified; therefore, it is a philosophical and not an emperical scientific argument. Penrose notes, "It tends to be evoked by theorists whenever they do not have a good enough theory to explain the observed facts." From my point of view, the Anthropic Principle is no explanation and tantamount to saying, "If things were different they would be different."

Guillermo Gonzalez and Jay Richards say the logical extrapolation of the Anthropic Principle, as it concerns the fine-tuned universe, would be interpreting the requirement for complex life is so because we have "selected it" by our presence as observers. As they point out, however, applying the Anthropic Principle to the fine tuning of the universe is far from a sufficient explanation and it is really no explanation at all.

Gonzalez and Richards write, "What is surprising is not that we observe a habitable universe, but that a habitable universe is, so far as we know the only one that exists." They give a familiar analogy of a popular story about a firing squad. Let's say in the past, an Army intelligence officer was captured by the Nazi SS in World War II and sentenced to die by firing squad. The SS officer employs 50 sharp shooters to take their position 10 feet away and fire at the American. After firing, none of them has hit their target but instead we see 50 bullet holes making a perfect outline of the intelligence officer behind him against the firing squad wall. Using an Anthropic Argument, the lead Nazi officer responds, "I suppose I shouldn't be surprised to see this. If the sharp shooters hadn't missed I wouldn't be here to observe it." I think if this really happened we would believe either the Nazi officer has lost his intelligence or his mind. We would clearly be looking for a better cause, like maybe the shooters intentionally missed their mark. Now, given this scenario enough times maybe it would occur just as it did. This is analogous to the many worlds hypothesis that many anthropic adherents now believe to be the case since it is somewhat absurd that all the circumstances necessary for complex life could have occurred by the production of only one universe. Our existence then to them becomes a "selection effect."

But, of course, there is a big difference between the Anthropic Principle and intelligent design. First, it is not obvious there are infinite numbers of universes and certainly there is no evidence for it except for the fact that the fine-tuning contradicts the Copernican Principle, as pointed out in *The Privileged Planet*. Astronomers Fred Adams and Greg Laughlin suggest the Anthropic Principle and hence the Copernican Principle can be extended to include that not only is our solar system not special, but neither is our universe!

Here is the problem in a nutshell. No matter how obvious the evidence is for an intelligent designer, naturalists will never accept it. Science should be the search for the best explanation, not for the most naturalistic one, and if acceptance of the Anthropic Principle is their only consideration that seems to be very irresponsible to me. The Anthropic Principle offers no causality, but intelligent design does.

The Copernican Principle has failed. The universe is fine-tuned, and our planet is uniquely placed in it for life to exist and to allow for discovery; and the best explanation for this is a creator who I believe to be the God of the Bible. Humankind has no problem discerning design when our faculties are fully functioning. The whole concept of the search for extraterrestrial intelligence is looking for patterns coming from the cosmos.

I am reminded of the great science fiction novel and movie by Arthur C. Clarke, *2001: A Space Odyssey*. In it, scientists excavating the moon find a perfectly rectangular "monolith" with the obvious implication being someone had placed it there. Nothing with measurements this precise could have occurred naturally; someone of intelligence must have created it and

2001: A Space Odyssey

put it on the moon. Why is intelligence so easily discerned in a case like this, yet ignored by scientists when design is even more obvious in the natural world?

It is a mistake, however, to assume the design argument is based only on calculations of probabilities or complexities. Complex life in our universe may be very rare but that in and of itself does not prove an intelligent designer. As we have stated before, it is pattern and a tight pattern that suggests a designer, a concept delineated by William Dembski. Gonzalez and Richards say, "It's the presence of a telling pattern, a pattern we have some reason to associate with intelligence agency, and not simply improbabilities that cause us to believe there is something behind all the fine tuning." From Gonzalez's and Richards's point of view, the fact that we can discover this fine tuning is even more evidence for an intelligent designer.

I believe the fine-tuned universe can be compelling evidence when trying to convert the gainsayer. I believe this is exactly what is alluded to by Paul in Romans 1:20, "For since the creation of the world, His invisible attributes, His eternal power and divine nature have been clearly seen, being understood through what has been made, so that we are without excuse." Probably the most gifted scientist in the last 100 years (or for that matter the past 1,000 years) put it this way in 1929,

> We are in the position of a little child entering a huge library filled with books in many different languages. The child knows someone must have written these books. It does not know how. It does not understand the languages in which they are written. The child dimly suspects a mysterious order in the arrangements of the books but doesn't know what it is. That it seems to me, is the attitude of even the most intelligent beings toward God. We see a universe marvelously arranged and obeying certain laws but only dimly understand those laws. Our limited minds cannot grasp the mysterious force that moves the constellations (Albert Einstein).

Chapter 18

How Old Is the Earth?

Earlier we observed evidence for an old universe. These evidences included the abundance of lite elements, the expansion of the universe, and the presence of cosmic background radiation. We also looked at the light-travel-time dilemma, and found the white hole cosmology of Russell Humphreys to at least give a plausible scientific answer from a young earth creationist point of view. There are, however, other evidences or contradictions or enigmas that seem to point to a younger universe.

One such example is seen in the earth-moon system. George Darwin, son of Charles, discovered years ago that the moon was slowly receding from the earth. His finding has been confirmed and indeed the moon is moving away from the earth at a rate of 4 cm per year, and is increasing its speed by 0.0016 seconds per century. Although this is a very slow rate, over time, there would be significant accumulation. If we fix the modern rate of recession and extrapolate it into the past, we realize the earth and moon would have been in contact with each other 1.3 billion years ago, or approximately one-third of the supposed age of the earth. At a billion years ago, the moon would still have been close to the earth and would have produced massive tides and no one really believes this to be the case. Now this does not prove a young earth but it does not disallow for it either, as studies by evolutionists have proven, the tides of the moon have always been steady which would not have been the case if the earth-moon system was 4.5 billion years old. They must theorize, "Some large event" as Danny Faulkner puts it, must have occurred more than a billion years ago for which they have no evidence.

The sun also provides evidence for a much younger solar system. As we have noted, the sun appears to get its energy from thermonuclear fusion of hydrogen into helium at its core. Theoretically, the sun has enough nuclear power to burn for 10 billion years and if it is 4.5 billion years old, then it must have slowly altered its core thus increasing the nuclear fusion rate brightening the sun. Calculations indicate that the sun should be 40% brighter today than when it was formed and also 30% brighter than when life was supposed to have arrived on the earth 3.5 billion years ago. The implication is obvious as either the earth would have to be much younger or it would have started out at severely lower temperatures than we see today, and there again is simply no evidence for that. So what is the naturalist's explanation? They assume the earth had much more greenhouse gases in the past than it has now and as the sun brightened these gasses were diminished. Danny Faulkner queries, "How two completely unrelated processes could have evolved in exactly compensating ways for billions of years is amazing."

Comets pose a problem for naturalists as well. Comets are fragile and are lost to collisions with planets, a phenomenon witnessed in 1994 with Jupiter. Also comets are ejected by gravitational forces and even wear out after a while. As a result of this, we can estimate the upper limit for the time comets orbit the earth. According to Faulkner, comets should have vanished after "a few tens of millions of years" ago. This obviously would rule out a solar system billions of years old. Of course, astronomers are aware of this; therefore, they propose hypothetical comet belts from which short period and long period comets originate, the Kuiper Belt and the Oort Cloud. The point is the Oort Cloud has never been detected and given its hypothetical distance from us likely never will be. Thus this is speculation and is not really scientific at all.

So how about the earth? Are there evidences that support a young earth, or does everything except a literal interpretation of Genesis point to a very old earth compatible with an age of 4.5 billion years? How one views the evidence depends to some extent on a person's point of view. If one cannot accept a supernatural cause, he will see only the evidences indicating an old earth. If, on the other hand, a person can believe in an intelligent designer, the evidences can certainly support that view as well.

Several years ago, my wife and I took a trip to Arizona and on that trip we went on a guided tour of the Grand Canyon, much like many people in the world have done. On this tour, our young guide pointed out in some detail the rock formations and how this great canyon was formed over the course of millions of years. He explained how the various sedimentary rock deposits were laid down and how the Colorado River had carved out this magnificent gorge. Of course I was not really in a position to argue with this apparently bright young man but it occurred to me that the vast majority of people believe like he does. After all, we are told over and over again by scientists, who are much more educated in these subjects than we are, that the evidence is conclusive; the earth is very, very old. But is that really true, and is it a closed deal concerning the age of the earth? Clearly, if Genesis is true and if the days of the creation are literal then the earth cannot be billions of years old. The Bible claims to be the inspired word of God (2 Tim. 2:16; 2 Pet. 1:21); and Jesus seems to have taken the Genesis account as literal (Mark 10:6-10). Given that man was created on the 6th day and observing the genealogies in the book of Genesis, the earth cannot be more than at most about 10,000 years old and some say 6,000. The only way this cannot be true for one who believes the Bible is either to treat the days of Genesis as non-literal days, but in a figurative way allowing for vast amounts of time, or to view the days in a relativistic time from the perception of the creator.

You might ask why even worry about it; the Bible is the word of God, I believe it, and that is all that really matters. This is certainly true enough; but how do you convince the non-believer, and does it even really matter how old the earth is anyway?

I would say the syntax of the book of Genesis indicates the days were days as we view them today and it potentially undermines God's authority to believe otherwise. Besides all of this, there is ample scientific evidence to indicate a young earth, and it is to this we now turn. Also, if a straight forward account of Genesis cannot be viewed as literal, where would you put the reigns on skepticism? Every other miracle in the Bible could be viewed as non-scientific according to methodological naturalism.

Uniformitarianism Vs. Catastrophism

Cosmology can be roughly divided by two major world views, uniformitarianism and catastrophism. In centuries gone by, catastrophism prevailed; but in the 1800's when the writings of lawyer-turned geologist, Charles Lyell, were published, uniformitarianism began to be accepted. Simply stated, uniformitarianism abandoned the belief in the biblical flood, instead espousing a concept that natural processes through time proceed at rates seen today. The mantra for uniformitarians is "the present is the key to the past." So when we observe the present rate of plate tectonics, the present rate of river erosions, and the present rate of radioactive decay, these processes must take millions of years to drastically change the topography of the earth. Certainly, this concept is unprovable but it seems to ignore how catastrophic events have changed the earth in the past and for that matter how they continue to shape our planet today. Interestingly, the Apostle Peter predicted this 2000 years ago. Look at 2 Peter 3 where the apostle was addressing the skeptics of the time. In verses 4-6 he quotes, "'Where is the promise of His coming? For since the fathers fell asleep, all things continue as they were from the beginning of creation.' For this they willingly forget: that by the word of God the heavens were of old, and the earth standing out of water and in water, by which the world that then existed perished, being flooded with water."

Catastrophism holds that the world can be changed and molded by major cataclysmic events, the most important one being the global flood we read about in Genesis. We see some of these events even today. Earthquakes, floods, natural dam breaks (which is a type of flood which probably formed the Grand Canyon from ancestral lakes to the northwest, as admitted by secular geologist now), volcanoes, hurricanes, and tornadoes can change geological features very rapidly. Even canyons and valleys can be created very rapidly, an example being seen in the Colorado River which does not erode its channels. Only a rapid catastrophic erosion on a giant scale could have carved out the Grand Canyon. We also see the effects of the eruption of Mount St. Helens and how it has significantly changed the landscape and its surroundings. In fact, at 1/40 scale, a version of the Grand Canyon was formed on the north side of Mount St. Helens by a mud slide. No catastrophic event ever witnessed by man could convey how much the flood of Genesis would have changed the earth.

The late John Morris, who held a Ph.D. in geologic engineering from the University of Oklahoma and was a department head of engineering at Virginia Tech University said, "This is indeed a wonderful time to be a young-Earth creationist because so much information is now available that confirms our understanding of scripture." He also points out in his book, *The Young Earth*, that many geologists are beginning to understand the significance of catastrophism and more and more seem to be embracing its concept. For example, Dr. Derek Ager, former president of the British Geologists' Association, has said, "The hurricane, the flood or tsunami may do more in an hour or a day than the ordinary processes of nature have achieved in a thousand years. ... In other words, the history of any one part of the Earth, like the life of a soldier, consists of long periods of boredom and short periods of terror."

This new breed of geologists are referred to as neo-catastrophists, and they believe nearly all the rock materials of Earth were laid down rapidly as sediments by catastrophic events. In fact, modern secular geology has recanted the philosophy that all geological formations have been created by uniformitarianism and all modern geology curricula includes at least some form of catastrophism in order to explain the Earth's geological features. Now they would claim much time would elapse between these events and these layers of sediment, but they have no proof of this whatsoever. In fact, as Dr. Morris points out, "The evidence for time is the lack of physical evidence." So let's turn now to the evidence for rapid deposition of geologic strata.

Evidence for Rapid Deposition of Geologic Strata

There are several features geologically that indicate rock formations can and have been laid down relatively rapidly. First, there are surface features. What do we mean by surface features? Surface features are those features seen on the top of a bed formation that would of necessity have to be deposited rapidly. These would include: ripple marks, rain drop impressions and animal tracks. Ripple marks would occur on a beach when tides recede, rain drops might actually be blisters formed by air bubbles escaped from rapidly deposited sediments and animal tracks would all be formed by soft surfaces or they would not be formed at all, and would be very fragile and unable to last for very long. There are many examples throughout the world where these formations are "frozen" in solid rock and would have to be preserved very rapidly or else they would fade away by erosion. In Texas, a great example of this can be found in Glen Rose where many dinosaur tracks can be seen at the Paluxy riverbed.

Dr. John Morris also points to bioturbation as evidence for rapid rock formation. On or below any rock surface, whether on land or sea, one would expect abundant life to be present and leave its mark. Dr. Morris gives the example of hurricane Carla in his book, *The Young Earth,* and says that recognizable layers of sediment were laid down both on and off the shore in 1961. They found twenty years later that, "Life at the surface of this bed, both onshore and off had destroyed internal characters that had been formed by these catastrophic processes." If you compare this layer of structure to other rock layers throughout the world, there is a great similarity. Now you can never say how long these lower layers of sediment existed before the top layer but what you can say is "That it was less than the time for bioturbation to destroy sedimentary structure within the lower zone" and conclude "a relatively short time for the entire sequence," says Dr. Morris.

There seems to be an almost complete lack of soil layers in the geologic column, and this also is consistent from a young earth perspective. Soil is needed for life, which is obvious, and should take a long time to form from a uniformitarian point of view. When land submerges beneath the sea some of the soil should be covered by the ensuing sedimentation, yet we don't see this commonly in the geologic column. Almost always the geologic column is one of rocks. Geologists have no good reason for this but a good explanation is that the time to produce the soil never happened!

Yet another peculiar finding for the uniformitarian geologists is the finding frequently of two totally different rock types lying one on top of the other with a "knife-edge" being between them. This can be seen in the Grand Canyon where the hermit shale lies below the whitish Coconino sandstone.

The Grand Canyon

The hermit shale contains many index fossils and the Coconino sandstone is believed to be formed by large sand dunes (or in other words, desert) which would have required a storm of unprecedented magnitude to move said sand. Dr. Morris notes, "If this sand were to be deposited over the hermit shale and produce a desert, as most uniformitarian geologists believe, significant erosions from such upheaval should have occurred and would have caused a complete flat surface on the shale which we do not see in this case." "The point is, the existence of the sharp, knife-edge contact between those two formations argues against the passage of long periods of time between their depositions," writes Dr. Morris. On the other hand, if there was a catastrophic event shifting this sand, we would expect to see exactly what we see – that being a very abrupt change in strata.

Possibly even a bigger enigma for the uniformitarian geologist are something known as polystrate fossils. Polystrate fossils are fossils, frequently of trees, that penetrate through overlying formations; they are frequently found in underground coal mines and are referred to as kettles. Dr. James

Hodges, in his well-documented book, *Creation vs. Evolution,* cites many examples of polystrates. Polystrates are found in Nova Scotia that extend through shale, sandstone and another layer of sandstone. Similar findings occur in Virginia and West Virginia, some of which Hodges examined personally. What is the implication of this? Simply that strata are deposited rapidly or how could trees traverse them? Only something very catastrophic could explain this.

Other evidences of rapid strata formation include cross-bedding on sandstone seen as parallel bedding lines or lami-

Polystrate Fossil

nations at oblique angles found in many places like the Coconino Formation in the Grand Canyon, cyclothems (repeated cycles of similar sequences of strata extending vertically) again found in many places worldwide, upright fossils occurring in Germany, Alaska, Pennsylvania, Great Britain, and other places, and turbidity current deposits. This latter is especially interesting.

Turbidity current deposits are essentially formed by underwater avalanches and graded layers of deposits can be formed in some cases in as little as a few hours. It was previously thought that these layers of stony conglomerates with cross-bedded sandstone on top, which in turn is topped off by siltstone, took years to develop, in the neighborhood of 1,000 years per 1 vertical foot of shale. This, however, is not true as new data indicates they can be formed very quickly. Dr. Hodges cites several examples of this phenomenon. In 1929, a great earthquake occurred on the Grand Banks of Nova Scotia which cut 12 transoceanic cables. By figuring the speed of the avalanche, scientists calculated the avalanche advanced at 63 miles per hour. The turbulence created debris that flowed 500 miles from the source. The deposits from this were 300 miles long and from 20 to 100 miles wide and up to 3 feet thick. Some 23 years later, when reexamining these deposits, previously called "flysch," geologists realized just how quickly all this could occur.

Turbidity currents also transform river, shore, and continental shelf deposits into the ocean depths. These flows can produce gigantic submarine canyons, some even rivaling the Grand Canyon. Again, these represent rapid events in many, if not most, cases. Harold Coffin, former professor of geology, has written, "Turbidity currents with the resulting turbidities have forced a major change in the interpretation of many sediments from slow gradual accumulations to sudden, almost instantaneous depositions." Turbidity deposits are found all over the globe and from all supposed geologic periods from the Pre-Cambrian to the Pleistocene. Dr. Hodges has concluded, "A world-wide flood would provide optimum conditions for large mudslides, particularly in the recessive stages as ocean basins settled and mountain ranges rose."

A lot can be learned by recent catastrophic events which show evidence of continual depositions laid down very rapidly. The prime example I am referring to is the explosion of Mount St. Helens that occurred on May 18, 1980. For those of us alive then, we can remember vividly the television pictures coming from this massive volcanic eruption as 150 square miles of forest north of the mountain were immediately wiped out. Within minutes, some 4 million logs began floating into nearby Spirit Lake along with tons of organic materials such as ash and parts of trees. A few years later, these organic materials such as tree bark and woody parts that sank to the bottom of the lake were recovered and looked a lot like peat, which if buried and cooked would become coal.

Also interestingly, many of the logs floating in Spirit Lake started floating upright when they became waterlogged. The upright "trees" will

Spirit Lake

eventually be buried in more sediment as it accumulates, which will further result in an upright "polystrate" position. These trees, waterlogged at their root bulb end, eventually sank to the bottom of Spirit Lake, with current estimates placing 20,000+ logs partially buried in sediment at the bottom of the lake. Finally, when looking at the sediment accumulation that occurred after the eruption, it looks amazingly just like the strata that we are told by uniformitarian geologists takes millions of years to form which in this case is, of course, not so. Steve Austin conducted landmark work, diving into Spirit Lake in order to study the piles of organic sediment and even witnessed trees falling to the bottom of the lake.

With all these things said, it would be a mistake to say we can "prove" the accuracy of the Bible by looking at the geologic strata. What we can say, though, with some definity is that the geologic strata is compatible with the biblical account. Furthermore, we can also say much of the geologic data is incompatible with an old earth scenario. Dr. Morris has written, "We cannot prove the Bible from looking at geology, nor do we try. We accept scripture by faith, but insist that if the Bible is really true, then geologic evidence must support it – and indeed it does."

Radiometric Dating

It is believed by most scientists that the earth is 4.54 billion years old. This number is purportedly arrived at using radiometric dating, also called radioisotope dating; and virtually every textbook or media journal on science refers to this dating system as the most accurate, absolute, method of dating our earth. For a creationist, and one who believes in the literal interpretation of the Bible, one must question how can the earth be 4.54 billion years old when the Genesis account requires a much shorter timeframe? To answer this question we have the obligation of considering and examining radiometric dating head-on.

The first thing to understand about radiometric dating is that its validity relies on several assumptions for it to be true. We will look at those assumptions subsequently, but before we do that, we must briefly describe the science about radiometric dating.

Radiometric dating is used to date igneous rock and not sedimentary. Igneous rock is rock that at one time was molten or hot and then cools (metamorphic). This cooling period then resets the minerals in the rock and starts the dating calculation. All matter in the universe is made up of atoms. The word atom is derived from two Greek words meaning undivided or indivisible; thus the atom is the smallest unit in which an element can be identified. All elements are made of protons, neutrons, and electrons. The nucleus of an atom contains protons and neutrons (which are, incidentally, both composed of quarks and gluons), with the electrons making up the outer core. The number of protons, which is referred to as the atomic number, identifies the element. In other words, hydrogen has one proton, helium two protons and carbon six protons, etc. The number of protons and neutrons in the nucleus make up the atomic weight (electrons are too light to add to the atomic weight). Atoms can have more than one atomic weight and when this is the case they are referred to as isotopes. For example, ^{14}C is the carbon atom with six protons but eight neutrons. In its stable non-isotopic form, ^{12}C, carbon has six protons and six neutrons. Isotopes frequently are unstable because they lack the energy to "hold on" to the extra neutrons and, as a result, can "spit out" these particles and decay into other atoms. This, in essence, is the basis of radiometric or radioisotope dating. When an atom of an element decays into the atom of another element by loosing subatomic particles, referred to as alpha decay, knowing the half-life of this process we can ostensibly calculate the age of rocks by determining the amount of "parent element" vs. the amount of "daughter element," parent being the original isotope and daughter being the decayed atom that results. In other words, by measuring the amount of daughter element in a rock and knowing the half-life, or the decay rate of the element, rock dating is allegedly accomplished. Note, however, that the supposed dates are simply an interpretation imposed on the rocks by scientists. Why? Radiometric dating in and of itself, says nothing about age and is only a measurement of the chemical composition of the rock. The "ages" must be inferred upon the rock by uniformitarian hypotheis, insisting that the Earth must be billions of years old.

Common elements that are used in radioisotope dating are uranium, potassium, and carbon. Uranium with an atomic weight of 238 can decay into lead with an atomic weight of 206 and this is calculated to take 4.47 billion years; potassium decays into argon taking 1.25 billion years only if the decay rate is assessed at today's slow and steady uniformitarian conditions, and ^{14}C can decay into nitrogen-14 which takes approximately 5,730 years again when based on slow, steady decay rates we see today.

For radiometric dating to be accurate, at least three critical assumptions must be made. First, the initial condition of the rock (that is the condition of the rock at its creation) needs to be known. Secondly, the amount of parent or daughter elements in the sample cannot have been altered in the process by any means other than decay. Thirdly, the decaying rate of the parent must have remained constant during this whole time.

Frequently, an hourglass is used to illustrate the dating method as well as the assumptions thereof. For example, if you see an hourglass that is half full you would understand that 30 minutes has passed by. This would be a valid assumption; however, it could be affected if we did not know how much sand was at the bottom of the hourglass to begin with, or whether any sand had been added or taken away, or if the sand had been continually falling at the same rate.

In the case of radiometric dating, there is possibly even a fourth assumption that John Morris describes in his book, *The Young Earth*, and that assumption is that the earth is at least old enough for the present amount of radioactive isotope. Another way of stating this is that radiometric dating assumes uniformity and assumes an old earth in its very presumption.

So, does radiometric dating really work and is it accurate? By accurate, I mean, does it really date the true age of the rocks being tested? To be clear, the actual measuring of the isotopes in the rock is accurate. Deducing the age is another thing altogether.

In 1997, a group of creation scientists came together to study radiometric dating by determining its consistency and accuracy in measuring the age of rocks. These scientists were all extremely well educated, with Ph.D.'s in physics, geophysics, geology, and atmospheric sciences. What they have found are many inconsistencies in radiometric dating, which should not be the case if these methods were reliable.

For example, Steven Austin, a geologist and a member of the RATE team, which stands for Radioisotopes and the Age of the Earth, dated the rocks from Mt. St. Helens using potassium-argon. We know the precise date when these rocks were formed, as the eruption of Mt. St. Helens was in 1980. Using potassium-argon dating, the age of the rocks from Mt. St. Helens was given as anywhere from 0.5 million years to 2.8 million years, clearly destroying assumption number one. This was also performed on rocks from Mt. Ngauruhoe in the north island of New Zealand, which erupted in the years 1945, 1954 and 1975. When these rocks were sent to three separate professional labs, the ages were given from 0.27 million to 3.5 million years old. You can see the problem: if you can't get accurate measurements on "known rocks," how can you trust the measurement of the unknown?

To counteract this, isochron dating, which involves analyzing four or more samples from the same rock unit, was instituted by scientists to eliminate the assumptions of starting conditions within rocks. Unfortunately, even with isochron dating wide variations were found, in some cases as much as 500 million years when analyzing samples from the Cardenas Basalt, considered the oldest strata in Eastern Grand Canyon. What explanations can be given for this discordance in age? First, there could be mixing of isotopes between volcanic flow but this can be eliminated by determining whether such mixing occurred, which can be easily done. Secondly, it's possible that minerals could have solidified at different times but there is no evidence that lava cools and solidifies in the same place at incredibly slow paces; thus this cause for discordance can be eliminated as well. The more likely explanation of the discordance is that the decay rates may have been different in the past than they are today, an explanation that most scientists are unwilling to consider.

Much of the work done by the RATE Group has been in providing evidence that radioactive decay actually supports a young earth and decay rates in the past must have been different than they are today.

One example of this is the amount of helium found in granite in the form of zircon crystals which contain radioactive uranium (^{238}U) decaying into lead (^{206}Pb). During this decay, eight helium atoms are formed in micron out of the zircon and granite rapidly. The helium in the zircon crystals should have migrated out by our present time but instead we see lots of helium. Why so much helium? Decay rates at one time were greatly accelerated. In other words, helium was being produced faster than it could escape thus leaving lots of helium still in the rocks. The RATE's conclusion was that billions of years of decay occurred within 4,000 to 8,000 years, suggesting this occurred during the creation day or during the Flood. "Based on the measured helium retention, a statistical analysis gives an estimated age for the zircons of 6,000 plus or minus 2,000 years. This agrees with the literal biblical history and is about 250,000 times shorter than the conventional age of 1.5 billion years for zircons. The conclusion is that helium diffusion data strongly supports the young earth view of history" (DeYoung from his book, *Thousands ... Not Billions*).

Another discovery indicating decay rates were different in the past is that of polonium halos that are found in granite rocks. Polonium isotopes producing "halos" in granite rocks are daughter products of ^{238}U isotope decaying to ^{206}Pb. As we have seen, the half-life of this process is an alleged 4.47 billion years. The cooling effect of granite over millions of years would have produced significant lead; therefore, allowing geologists to assume a very old age for these granite rocks. Polonium, on the other hand, in its isotopic forms of 210, 214, and 218 has half-lives of only 138 days, 0.00016 seconds, and 3.1 days, respectively. By looking at biotite crystals containing halos in the granite, no uranium is found in them so polonium must have been in the biotite as the granite rock cooled. James Hodges states in his book, *Creation vs. Evolution*, "The presence of polonium halos in the biotite crystals means that the time between the embedding of the polonium and the formation of these crystals had to be less than the youngest half-life of the enclosed polonium isotope (138 days)." Hodges alludes to research performed by R.V. Gentry. Gentry, a research specialist for thirteen years at Oakridge National Laboratory, observed that if radiohalo-bearing granite had cooled gradually over hundreds of millions of years, "Polonium halos could not possibly have formed because all the polonium would have decayed soon after it was synthesized." Gentry concluded, "If these results are correct, then the fundamental postulate of modern geochronology, i.e. the assumption of the uniform radioactive decay rate, is called into question; and indeed the entire framework of modern geology is open to reinterpretation using different premises."

One other bit of evidence for a non-uniform decay rate in the past is given by the RATE group as it concerns fission tracks which provide a physical record of nuclear decay within a crystal. These tracks are produced by the splitting of very heavy, unstable atoms such as ^{238}U into smaller atoms like palladium-119 (^{119}Pd). The density of these fission tracks in crystals has been used by geologists to date volcanic rock. The RATE team wished to determine if the fission track density (thus the fission track ages) match the quantities of nuclear decay measured as the radioisotope age of the same rocks. What they found was, in fact they did not. For example, a discordance between 75 million vs. 169 million years was found in the rocks from Cambrian geologic formations. Further discordance was found in rocks measured from the upper Jurassic system with dates ranging from 88 million years for the youngest rock to 1,036 million years plus or minus 310 for the oldest. Hodges concludes, "The fission track density agreed with the excessive helium retained in crystals embedded in Pre-Cambrian granodiorite and with the abundance of parentless polonium halos along with ^{238}U halos preserved in granite masses and coolified wood as evidence for accelerated nuclear decay." So how do geologists deal with evidence for a young earth? In essence, scientists have philosophically ruled out a young earth, viewing it as unscientific and certainly realizing it cannot support evolutionary theory. Therefore, any evidences to the contrary are either given *ad hoc* explanations or considered to be isolated anomalies which do not fit in the world paradigm of an old earth (translation: they ignore the evidence).

The evidence mentioned thus far should not leave the impression that radioisotope dating has been overturned. John Morris says, "It has been called into question, flaws in its foundation exposed, and its results shown to be inconsistent. In short, it is in trouble, but it's still a very formidable concept in the minds of many." Research by creationists such as the RATE group will continue and much is still to be done.

^{14}C Dating

We have noted radioisotope dating is used by scientists in determining the age of rocks but how about determining the age of fossils? This is where ^{14}C dating comes in. Carbon exists in at least three forms. ^{12}C, the more common form and the non-isotopic form, contains an atomic weight of 12 and an atomic number of 6; that is 6 protons and 6 neutrons. Isotopes ^{13}C and ^{14}C exist, the most unstable being ^{14}C, which has 6 protons but 8 neutrons. ^{14}C is radioactive and will decay through beta decay where a neutron in ^{14}C is converted into a proton thus changing into nitrogen which has an atomic weight of 14 with 7 protons and 7 neutrons. ^{14}C itself is made by cosmic radiation with cosmic rays colliding with atoms in the atmosphere and neutrons that come from this collision with nitrogen to convert them into the ^{14}C isotope. ^{14}C then combines with oxygen to become carbon dioxide, CO_2, which gets incorporated into plants, animals, and all living things; thus all living things contain the same ratio of ^{14}C and ^{12}C, which is basically in the air we breathe.

A key thing to remember is, as long as an organism is alive it will contain the same ratio of ^{14}C and ^{12}C; that ratio being one ^{14}C atom to every one trillion ^{12}C atoms. At death, this ratio will begin to change as the ^{14}C atom is unstable; thus, the smaller the ratio the longer the organism has been dead. Since the half-life of a carbon atom is relatively short, being 5,730 years, the carbon dating method can only be used to date fossils that are no more than 50,000 to 60,000 years old since by that time all the ^{14}C will have been decayed. As you can see with that assumption then, ^{14}C dating actually gives more evidence for a young earth than an old one. And, using accelerated mass spectrometry, ^{14}C can be very precisely measured.

^{14}C dating, just as isotopic radiodating, must also assume several things in order for it to be accurate. For example, assumption must be made, again, as to how fast ^{14}C decay has been in the past and what was the starting amount of ^{14}C in the organism when it died. An even more critical assumption is the ratio of ^{14}C to ^{12}C in the atmosphere. Has it always been the same as it is today, as a difference would make measurements invalid? In *The New*

Answers book edited by Ken Ham, Mike Riddle states that the founder of the ^{14}C dating method, Dr. Willard Libby, assumed this ratio was constant; however, his reasoning was based on an evolutionary concept more than a scientific one. Riddle points out that Libby's original work showed that the atmosphere did not appear to be at equilibrium; in that if the earth started with no ^{14}C in the atmosphere it would take up to 30,000 years to build up a steady state (equilibrium). As Riddle states, "Dr. Libby chose to ignore this discrepancy (non-equilibrium state), and he attributed it to experimental error. However, the discrepancy has turned out to be very real. The rate of ^{14}C, ^{12}C is not constant." Obviously, then, if the ^{14}C is still not at equilibrium the earth must be very young.

Other factors that affect the production of ^{14}C include the changing magnetic field. We know the magnetic field is decaying, or getting weaker. As the magnetic field weakens, more cosmic rays are able to reach the atmosphere. Thus, at an earlier time, less radiation would be around to produce ^{14}C in the atmosphere.

The flood recorded in Genesis may have also played a significant factor in the amount of carbon. The flood obviously destroyed a massive amount of plants and animals which in essence formed today's fuels of coal and oil, which in turn means that just prior to the flood the biosphere would have contained as much as 500 times more carbon in living organisms than today. This would thus dilute the amount of ^{14}C and cause the ratio to be much smaller than it is today. Riddle concludes, "When the flood is taken into account along with the decay of the magnetic field, it is reasonable to believe that the assumption of equilibrium is a false assumption."

Finally, the RATE group demonstrated the results of ^{14}C dating had serious problems for long geologic ages. They took samples from ten different coal layers representing different periods in the geologic column that were dated millions to hundreds of millions old by standard evolutionary estimates, and found all samples contain ^{14}C. Why is this significant? Again, the half-life of ^{14}C is relatively short and one would expect no detectable ^{14}C after 100,000 years. The more reasonable estimated age of these layers would have been 50,000 years but factoring in a more realistic pre-flood, $^{14}C/^{12}C$ ratio reduces this significantly more. The RATE group even looked at the amount of ^{14}C in diamonds. Diamonds are believed to be millions and millions, if not billions of years old and clearly should contain no ^{14}C; however, as you might have guessed ^{14}C was indeed found in twelve different diamond samples tested by the RATE group. What's the conclusion? The conclusion is ^{14}C can never be used to date objects more than 100,000 years old. How are dinosaurs and other organisms dated to be millions of years old? That involves the concept of fossil indexing which, as we have seen previously in our studies, involves circular reasoning. Radiometric dating is used to date rocks where fossils may be found, and thus the fossils in the rocks are given the same age. Furthermore, when a fossil is believed to have lived millions of years ago and another organism is found in association with that fossil, its age is given as the same. In conclusion, all radiometric dating methods are based upon assumptions of what occurred in the past which cannot be proven. When these assumptions, which are all biased towards an old earth, are removed or shown to be faulty, the results then support a biblical young earth. We do not have to be afraid of radiometric dating methods and in the case of ^{14}C, it is actually our friend and supports a young earth.

Chapter 19

Beginnings

"In the beginning God created the heavens and the earth." The beginning words of the Bible are arguably the most important words ever written. The writer of Genesis assumes God and makes no effort to prove it. It seems logical to me that, if there is a God, He most likely would have let that fact be known in some form or fashion. I believe He did, through natural revelation and special revelation – His book!

Through the ages, a number of books have claimed inspiration but nothing comes closer to proving it than the Bible. The Koran, the Hindu Vedas, the work of Confucius, and all the other so-called inspired works contain no predictive prophecies that have come to fruition like the Bible has. The Old Testament contains literally hundreds of prophecies that have come to pass. In Genesis, God promised to make of Abraham a great nation; and indeed a great nation did come about through Abraham and his wife, Sarah, in the birth of Isaac in spite of their advanced age (Gen. 12:2; Exod. 2:37) and, as a result a great nation, the Hebrew nation, was eventually formed and fulfilled the land promise during Joshua's days (Josh. 21:43, 45). Isaiah, prophesying many years before they occurred, foretold of things that would happen to Babylon, Moab, Syria, Ethiopia, and Egypt (Isa. 13 through 19). Daniel prophesied things concerning Babylon, Persia, Greece, and the Roman Empire, and they came true as well (Dan. 2). Jeremiah, Ezra, and Nehemiah all foretold of the remnant of Israel returning from captivity in seventy years, which of course occurred precisely as foretold. Isaiah even mentioned by name the king who would allow this return, Cyrus of Persia, 120 years before he was born!

Just the prophecies concerning Christ alone should be enough to convince anyone of the inspiration of the Bible. His virgin birth, the place of His birth, the establishment of His kingdom, the mechanism of His death, His resurrection, His ascension, and many more prophecies all came to pass and were prophesied hundreds of years before they occurred. Pete Stoner in Josh McDowell's excellent book, *The New Evidence That Demands a Verdict*, calculated the likelihood of 48 prophecies of Christ coming true by mere chance to be one in 10 to the 157th power or a one with 157 zeros after it. As you might surmise, this is a very, very, very large number. Just the fulfillment of eight prophecies alone coming true would be 10 to the 17th power which as Stoner illustrates would be the equivalent of finding one marked silver dollar in the state of Texas with unmarked silver dollars covering the entire state by two feet deep.

Consider this as well, the Bible was written over 1,500 years by more than forty writers, who came from various works of life including kings, military leaders, peasants, philosophers, fishermen, tax collectors, poets, musicians, statesmen, scholars, farmers, physicians, and shepherds, and was written in three languages, on three continents, and yet it presents a unified story of God's redemption of mankind. The unified theme throughout the book is man's salvation through an unwavering faith and obedience to His word with the one leading character throughout the book, God, being known through His son, Jesus Christ.

Lewis S. Chafer has said, "The Bible is not such a book man would write if he could or could write if he would. The Bible presents reality, not fantasy, with the good parts and with the bad." Even the "heroes" of the Bible, save for Christ, all had their flaws and sins, yet none of those were hidden in the text from the reader.

Finally, the historicity of characters and events of the Bible have been confirmed by non-biblical sources through archeological discoveries. There can be no doubt of the historical evidence of Christ's existence either. Several non-biblical writers such as Tacitus and Josephus spoke of Christ and Christianity, and there is historical evidence even of His resurrection. There is just too much evidence to ignore the inspiration of the Bible. God has chosen it as His vehicle to communicate to man and has been doing so for thousands of years now. His word has been confirmed by the miracles performed by the prophets and the apostles and of course Jesus Christ Himself, and further the Bible proclaims that no new revelation will be forthcoming (Jude 3).

If we believe the Bible is the word of God revealed through man, one must ask, why does it not explain everything? Why does it not describe all the processes of the universe such as gravity, electromagnetism, quantum physics, etc, etc? First and perhaps foremost we must understand the Bible is not a book of science, it is a spiritual book whose chief concern is the salvation of mankind. It should also be kept in mind that it was written in parts, some of which are over 4,000 years old, and yet it was written for all ages, ours included. Had it been written with scientific jargon, ancient man would have no capacity for understanding it, and no doubt even modern man's appreciation and understanding of science will have to be corrected at least in part, and therefore we might not be able to comprehend the processes even now. The great theoretical physicist of the 20th century, Richard Feynman, has told us science is the process of learning and then unlearning some part of the truth.

In the Beginning

The book of Genesis implies the beginning of everything man can observe. Literally, everything we see around us, the stars, the moon, the earth, and life all had a beginning and are elucidated in the first several chapters.

God, of course, had no beginning and we understand He transcends time and space, things as mortals we accept as self-evident. None the less, it is my belief that the Genesis record includes a beginning of all the fundamental laws and processes mankind can witness or discover, albeit, our understanding of much of God's creation remains only fragmentary. Light, which includes electromagnetism, matter, space, gravity and even time itself, had beginnings as did the core fundamental processes of the universe as we understand them now, relativity and quantum mechanics. Though some of these concepts are not precisely described, I believe they are certainly incorporated in the creation account.

The book of Genesis also deals with the beginning of order and complexity, neither of which would arrive spontaneously; they are always produced by prior cause. The book specifies the origins of such other things as marriage, evil, language, government, culture, nations, religion, and the Hebrew nation, but the scope of our study will be concerned only with the origin of the universe and our earth.

The first verse in the Bible then is one of the greatest statements in the history of humankind, and has profound implications as it is the beginning of "everything." It also makes several assumptions that merit our investigation. First, a beginning assumes a cause, and the only thing that existed before the beginning was God, which is the basic assumption of the first verse in Genesis. The creation event marks the start of the universe and time. The creation event was *ex nihilo* as described by Dr. Stephen Meyer, which means something came out of nothingness. Further, Dr. William Lang Craig, Ph.D., Th.D., states that the Kalam argument gives strong evidence for a creator. The Kalam argument, formulated by al-Ghazali (an Islamic philosopher), states, "Whatever begins to exist must have a cause." Since the universe began to exist, it therefore must have had a cause. Though some have argued the universe is eternal, Dr. Craig states "the scientific evidence has accumulated to the extent that atheists are finding it difficult to deny that the universe had a beginning." We will subsequently discuss these evidences at length, with their profound implications, but it must be stressed once again that what begins to exist must have had a cause. The universe then was created out of nothing as stated earlier and only a supreme being such as God could have created it. The writer of Hebrews 11:3 verifies this through the Holy Spirit stating, "Through faith we understand that the worlds were framed by the word of God, so that things which are seen were not made of things which do appear."

Even though metaphysically it seems obvious that something that begins must have a cause, atheists will still affirm, "Nothing came from nothing and for nothing." What is a proof then for a creator? To believe something came from nothing for nothing requires a philosophy that is "worse than believing in magic."

Naturalists of course would argue that if the universe had a beginning then God must, therefore, have had a beginning as well. They fail to understand, however, that God transcends time; therefore, he is beyond time. It is meaningless to speak about the beginning of God as God has no beginning and no end. This is very difficult, and possibly impossible, for mortal man to comprehend, yet true nonetheless.

So when did time originate? Back to Genesis 1, in the beginning. When the heavens and the earth were created, time started. Before that, there simply was. The fact that time had a beginning is clearly demonstrated in such passages as 2 Timothy 1:9, which states, "This grace was given us in Christ Jesus before the beginning of time," and in Titus 1:2, "The hope of eternal life which God, who does not lie, promised before the beginning of time." The fact that time has a beginning has serious implications. The cause of the universe, then, must be some entity operating in "a time dimension completely independent of and pre-existent of the time dimension of the Cosmos." Professor Hugh Ross, Ph.D., states, "The creator is transcendent, operating beyond the dimensional limits of the universe. God is not the universe itself nor is God contained within the universe. Pantheism and atheism do not square with the facts."

Time has certain qualities that God gave it that warrant further discussion, as well. First, it had a precise beginning, an instant of existence, which we will call "the Creation Event" and what scientists call "the Big Bang," which was first coined derisively by astronomer Fred Hoyle. Dr. Hugh Ross, a progressive creationist, has termed the Big Bang an unfortunate moniker, because it implies an explosion or bang or blast, which would yield disorder and destruction; yet what occurred according to some Big Bang theorists was a controlled, "carefully planned powerful release of matter, energy, space and time within the confines of very carefully fine-tuned physical constants or laws that govern their behavior and their actions."

The beginning of the universe heralded the beginning of all laws and constants as we know them. Dr. Morris points out, the word "heaven" in the first book of Genesis comes from the Hebrew word *shāmayim*, which is a plural noun and can mean heaven or heavens. It does not necessarily mean the stars, as stars and galaxies were created later. It most probably represents the concept of space. Furthermore, again in the first verse of Genesis, the word "earth" does not convey the meaning of the physical earth we inhabit at this time. The creation of earth would better be characterized as the creation of the basic elements of the universe as originally the earth had no form (Gen. 1:2). Genesis 1:1, therefore, speaks of the creation of the time-space-matter continuum and would include all physical laws of the universe that were created. These laws, as we best understand them now, would include special relativity, general relativity, and quantum mechanics. We understand that these laws may be modified at some time in the future but at least at this point in mankind's history, the universe seems to operate based upon these mechanisms.

For example, special relativity states, "All the physical laws of nature are the same for all observers in uniform motion." On the surface, this seems rather obvious (and thankfully so for us or else one could not eat a meal on a train or plane), but this law, if you will, has other interesting implications. For example, time is not absolute as stated before but depends on a frame of reference. In essence, this means time can be faster or slower depending on where you are and how fast you are traveling. To illustrate; clocks run slower in regions of higher gravity due to the "warping of the space-time grid," and time runs slower the closer one comes to traveling at the speed of light. One might argue this "time dilation" is about light or clocks but it is really about time itself.

To conceptualize time dilation, let's look at an example given by Stephen Hawking in his book, *A Brief History of Time*. Dr. Hawking says that relativity demonstrates that as one approaches the speed of light, time slows down. If we could build a spaceship that travels near the speed of light, let's say 90%, and if we sent an astronaut to a star one light year away and back, the astronaut would age approximately two years on his journey, as that is the amount of time it would take to get there and back in his time reference. Unfortunately, when he or she arrived back on earth they would encounter a whole new culture, as the earth would have aged well over 700 years. This phenomenon of time dilation is due to relativity and the unique properties of light, and has been experimentally demonstrated many times. Of course, currently mankind has no machines that even approach the speed of light, but Hawking believes that will not always be the case. What should be noted is this phenomenon of time dilation affects only time going forward as the arrow of time can never go backwards, which seems to be a fundamental law of physics. As mentioned previously, this quality of time is the part of the white hole cosmology discussed which can solve the light years dilemma.

Another prevailing physical reality that would have been created at the beginning of the universe is general relativity, first espoused by Albert Einstein in 1916. Einstein, in my estimation the greatest scientific mind of the 20th century, treated time as a fourth dimension, sometimes called the "entity of space and time." What is amazing to me is that Einstein postulated his theory of relativity at a very young age, in his early 20's, while working as

a patent clerk in Germany. These most profound theories, possibly the most important theories in human history, general relativity and special relativity, were all derived from thought experiments performed by Einstein, as Einstein did virtually no actual physical experimentation. Einstein would spend countless hours thinking about things you and I would just take for granted, like light and gravity. Einstein was not necessarily interested in the how so much as the why. In Einstein's general theory of relativity, gravity was thus explained. Gravity for Einstein was the bending of the space-time continuum, and evidence for this was even-

Albert Einstein

tually demonstrated by Sir Arthur Eddington in the eclipse of the moon that occurred in 1919, which has been further repeated many times. As a result of this, general relativity predicts the universe is either expanding or contracting and Edwin Hubble, in 1928, using the Doppler Effect which he discovered, proved that the universe was indeed expanding. This is, in fact, consistent with Psalm 104:2, which mentions the "stretching out of the heavens like a curtain." So far, no experiment has ever been discovered that has disproved either general or special relativity.

The other universal mechanism instituted at the creation event would have been quantum mechanics. Quantum mechanics defines how atomic and subatomic particles behave. It involves the uncertainty principle postulated by Werner Heisenberg, which states the velocity and location of a particle can never be absolutely determined at the same time. It also implies that the universe is not necessarily deterministic, and posits a wave-particle duality of electrons and indeed matter. This results in probability distributions delineated by Erwin Schrodinger in his famous wave equations. Much of quantum mechanics is intuitively difficult to grasp. For example, quantum theory states that electrons and even light have both a wave and particle duality associated with them. Yet even though quantum mechanics may not "feel right," quantum theory remains a very successful theory and is responsible for the technological advances that have occurred in the last fifty years.

Some have argued that quantum theory and relativity violate biblical principles; but do they? The Copenhagen Interpretation of quantum theory, first put forth by Danish physicist, Niels Bohr, attempted to bring the physical ideas of quantum theory into a more philosophical realm. Non-determinationalism and the uncertainty principle lead some to believe that reality was in question. Even Einstein did not like the uncertainty principle and famously stated, "God does not play with dice" as an expresion of this dislike. Yet, the Copenhagen interpretation of quantum theory is, in fact, a perversion of the theory. Some have also argued that relativity is equated to moral relativism but, once again, this is a distortion of relativity. The physical world should not be equated to the moral or spiritual world.

What about string theory? String theory is an attempt to "explain everything" and combine relativity with quantum theory. The mathematics that govern relativity and the mathematics that are involved with quantum theory simply don't jibe together. It would appear these bastions of physics are mutually exclusive. Nonetheless, relativity does an excellent job of explaining the universe, and quantum theory an excellent job of explaining the subatomic world. As a result, physicists for decades have been looking for the great unifying theory, the one theory that would merge relativity and quantum mechanics and explain "everything," and the current string theory might just do that.

String theory is a very difficult theory to grasp, but in essence states that all matter consists of very tiny strings that vibrate at different frequencies. For string theory to be valid, an additional six dimensions "tightly curled up in the particles themselves must exist." Today, most physicists consider string theory a promising theory and it could lead to new technologies; but string theory cannot be documented at this time, because extremely high energies are required to accelerate particles to expose the "six dimensions," and string theory and such particle accelerates simply do not exist at this time.

What's the conclusion then? First, God, who transcends time, has created time, space, and matter in the beginning. All physical laws were created at that time and the current scientific thought relies on two pillars; quantum mechanics and general relativity. Both have tremendous experimental support and Christians need not view these concepts with suspicion. They are, however, very incomplete and one would expect that science would someday replace these theories with different ones, or at least modifications of them. As Christians, however, we realize any new theory should be interpreted and guided by the Bible, as the Bible does not contradict science.

Genesis 1:2-3

After creation of space, time, and matter, Genesis 1:2-3 seems to imply a creation of the forces of energy required for our universe, in particular light in verse 3, "And God said, Let there be light, and there was light." Interestingly, this is the first point in the book of Genesis where God is recorded as speaking. The key question to ask is: What was the source of the light? We know the stars, the sun in particular, all the luminaries were not created until the fourth day so the question, once again, is: What and where was the light being generated from? Some have said that God was the light referring to 1 John 1:5 which states, "God is light and in him is no darkness." They also cite further references such as Revelation 21:23, which says, "And the city has no need of the sun or the moon to shine upon it, for the glory of God has illumined it, and its lamp is the Lamb." Furthermore, they reference Exodus 3:2 where God is speaking to Moses in the burning bush, which is obviously producing light, but fail to understand that the verse also says it was an angel of the Lord in the midst of the bush. Jesus is also referred to as light and refers to himself as light in 1 John 1:9; and Christians are even referred to as light in Matthew 5:24 when Christ calls us "the light of the world," which is a phrase Paul uses as well in Ephesians 5:8. Furthermore, in 1 Timothy 6:16 we read, "God dwells in unapproachable light," and in James 1:17 God is the "father of light" and finally Isaiah 45:7, "God formed light and darkness."

Clearly, light was yet to be produced and we are talking about physical light, I believe. God here created it. He was not the light; He created light, and to equate God with physical light shows either ignorance of what light is or the way light is used figuratively in the Bible. The aforementioned verses are clearly using light in its figurative sense. God is light in that through Him we are enlightened to His ways and to His greatness and majesty. In the same way Christ was the light of the world in John 1:9, and even Christians are looked upon as lights of the world by the way we live our lives and the teaching of the gospel. James 1:17 says, God is the father of light, but Isaiah indicates He was also even the father of darkness. The light mentioned in Genesis 1:3 is physical light.

This light at this point has energized the universe. Visible light is implied but all forms of light, I believe, are necessitated as well. So the question arises: What is light? We know now that light is electromagnetic waves. Light comes in both short waves, that are ultraviolet rays, and long waves, which are infrared light. So the range of light goes from very short waves, which are also cosmic rays and gamma rays, nonvisible to the human eye, all the way to infrared waves, even radio waves which, again, are not visible to the eye. The portion of light that is actually visible represents only a small fraction of all the wavelengths of light.

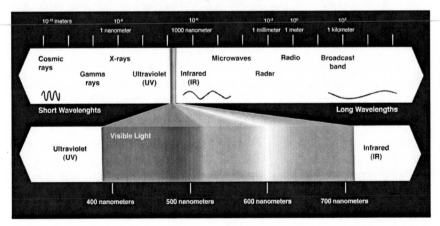

Another quality of light is that although it comes in waves, it also has particle quality to it as well, which are known as photons; this has been substantiated through scientific research. Light therefore has a duality similar to what we see in electrons, in that they have both a wave and mass function. Light also has a fixed speed, which is 299,792,538 meters per second; and nothing is known to exist in nature that exceeds this speed. So, how fast is light? Exceedingly fast; but nonetheless finite, an example being, it takes 8 minutes, 19 seconds for the sun's light to reach Earth. Light reflected off the moon, on the other hand, takes only 1.3 seconds to reach Earth. We also know that this light travels in discrete quanta, which is part of the origin of quantum theory. This energy has even been quantified with the equation: $E=HV$ where E is energy, H is Planck's constant, and V is the frequency of radiation. Einstein has quantified it with his famous equation of $E=MC^2$ where Energy is equal to Mass times the speed of light squared. So light obviously carries a significant source of energy. We know also that light is a requirement for life. Without light, there is no photosynthesis, no possibility of chemical reactions since light or electromagnetic waves are responsible for those reactions, and hence life could not exist.

Another quality of light is that visible light allows man to discover the universe. Without this spectrum of light and with our ability to see, discovering the universe through telescopes would be obviously impossible. Finally, another quality of light elucidated by Einstein is that its speed is always measured the same, no matter one's frame of reference. For example, if you were to measure light on Earth in a standing position you would get 299,792,538 meters per second; but you would get this same number if you measured the speed of light in an aircraft going 400 miles per hour. This seems to defy special relativity; but nonetheless it is an observable fact and leads to some interesting conclusions, and it forms the basis of Einstein's general relativity theory.

By the end of Genesis 1:3, according to Dr. Henry Morris, all the forces of the universe had been activated; those being electromagnetic force (light), gravitational force, and nuclear forces – both the strong and the weak forces. God was now ready to continue with his creation.

Genesis 1:4 and 5 states, "And God saw the light, that it was good: And God divided the light from the darkness. And God called the light day and the darkness he called night. And the evening and the morning were the first day." Here we see the cyclical process of the light and day arrangement beginning which seems to imply that the earth, though formless and void, still was beginning to somehow rotate on an axis, creating the evening and morning. We know that the Hebrew word for "day" (*yôm*) usually refers to a 24-hour day, although it sometimes can occasionally mean an indefinite time. For mankind, 24 hours is certainly indicative of a day which means a complete rotation of the earth. Using God's frame of reference, it is hard to be dogmatic to insist this first "day of creation" was a literal 24-hour time. Nonetheless, by the end of this first day, time, space, matter, gravitational forces, nuclear forces and the electromagnetic force were all active, it would appear.

Henceforth, the book of Genesis deals with the rest of God's creation culminating in the creation of man. Man was God's ultimate creation and it is clear mankind holds a special place in God's creation. Man alone was given an eternal spirit. In only man did God impart the ability to think abstractly and in only man does God demand worship and obedience to His will.

But man was also given an intellect and ability for discovery and through the passage of time we have uncovered more and more about God's wonderful creation and how it all seems to work. We now know the fundamental properties of life which we have discussed previously in this book. We also understand more precisely how gravity works through the great discoveries of Isaac Newton, how the planets orbit our solar system through Johannes Kepler, the fundamentals of electromagnetism from James Clarke Maxwell, and the unique qualities of light from Albert Einstein.

I believe God has given the ability to make these discoveries and observations to further demonstrate His glory and omniscience and strengthen our faith in Him. Appropriately applied science is not antagonistic toward Christianity but rather an ally when it is predicated on a divine creator.

Chapter 20

What Does It All Mean?

In the end every person holds to one of two world views, whether he can articulate it or not, and those two views are naturalism (materialism) and theism. The naturalist believes that everything in the universe is here for no apparent reason and came forth through naturalistic causes with no supernatural intervention. The theist believes a divine or supernatural force of infinite intellect created the universe by various processes, and those processes are still at work today. You may debate exactly how these processes were initiated, and how long they took to occur, or even if that is relevant or not. You may even argue about the processes themselves. You may also question if the divine has communicated to mankind, and whether the Bible or some other supposed spiritual work is His inspired word; but eventually you will have to decide which makes more sense, pure naturalism or intelligent design, and being agnostic, in my opinion, is really deferring to naturalism.

By intelligent design, of course, I am alluding to a creator; again, of superior supernatural intellect, the kind not seen on this earth. The intelligent design movement gives substantial support to this concept. I admire the intelligent design community and have quoted many times in this book many of its primary adherents. I believe Stephen Meyer, one of the leaders in the intelligent design community, to be a man of brilliance and integrity; but here is the problem I have with the movement as a whole. Using an analogy, they bring you to the trough, but the trough is empty of water. In other words, they point to God but they give no further guidance after that; nor do they even give due significance to the importance of the designer. I'm sure this is because the intelligent design scientists come from various beliefs and faiths; but I feel compelled to tell what I believe to be the whole truth.

So, matter has either always existed, sprang forth from nothingness, or was created by God; and I can think of no other alternatives.

For the naturalist, there is no realm of transcendence or the supernatural, and there is nothing but the natural. Carl Sagan, the now deceased astronomer, has encapsulated this naturalistic view from his PBS program *Cosmos* of years ago by saying, "The cosmos is all there is, or was, or ever will be." But arguing that something came from nothing is almost silly, for if there was nothing to begin with, you still have nothing after, which is in essence mathematical, 0 plus 0 equals 0! We have seen how some scientists now defer to a multi-universe explanation for everything, but this in reality is no explanation at all; not to mention, there is absolutely no shred of scientific evidence for it. Philosopher Richard Swinburne has argued, "To postulate a trillion-trillion other universes, rather than God, in order to explain the orderliness of our universe, seems the height of irrationality," as quoted in John Lennox's book, *God's Undertaker: Has Science Buried God?* Cosmologist Edward Harrison says, "We have two choices, blind chance requiring multitudes of universes or design that requires only one and the only evidence we have today is of one universe." Also from Lennox's book, Arno Penzias, previously mentioned co-discoverer of microwave background radiation, puts the argument another way: "Some people are uncomfortable with the purposely created world. To come up with things that contradict purpose, they tend to speculate about things they haven't seen."

The evidence for design is overwhelming to me, as I hope we have documented in the preceding chapters. God's fingerprints are seen by observing his finely tuned universe and this is what the apostle, Paul, spoke of in Romans 1:18-20, "For the wrath of God is revealed from heaven against all ungodliness and unrighteousness of men, who suppress the truth in unrighteousness, because that which is known about God is evident within them; for God made it evident to them. For since the creation of the world his invisible attributes, his eternal power and divine nature, have been clearly seen, being understood through what has been made, so that they are without excuse." The Psalmist in chapter 14, verse 1 has also told us, "The fool has said in his heart, There is no God."

We have also shown this fine tuning cannot be explained away by the anthropic principle; and the fact that our universe is rational and intelligible is a very strong argument for a designer. Albert Einstein once said, "The most incomprehensible thing about the universe is that it is comprehensible."

Again, Arno Penzias has said, "Astronomy leads us to a unique event, a universe which was created out of nothing, one with a very delicate balance needed to provide the exactly right conditions required to permit life, and one which has an underlying, one might say supernatural plan."

We have also discussed in this book how utterly impossible it would be for life to have evolved undirected from purely naturalistic causes, even if we could have somehow gotten the building blocks for life out of nothing. The DNA molecule, the basis of life, would be impossible to have self-assembled and requires a vast amount of information, the source of which cannot be explained by materialistic causes. Stephen Meyer has put it this way: "Self-organizational theorists explain well what does not need to be explained. What needs explaining is not the origin of order ... but the origin of information."

We have also noted that no matter how intriguing the Miller-Urey Experiment may seem, there is no indication that life began out of some primordial soup, as the evidence is that the earth's early atmosphere would be hostile to the formation of amino acids; not to mention the fact that the likelihood of its formation randomly of amino acids and thus proteins rises to the power 1:10 to the 60th. Furthermore, adding energy, the supposed mechanism of generating life, has been likened to "exploding a stick of dynamite under a pile of bricks and expecting it to form a house," so says Paul Davies, again from *God's Undertaker*.

We have also seen how irreducible complexity argues strongly against evolution. The physiology of vision, requiring forty complex biomolecules interacting with each other in picoseconds, is necessary for sight and must also be connected to a brain. That brain also needs to have evolved simultaneously. For vision to have been achieved by random natural causes is virtually impossible. We have seen how this same logic applies to other systems as well, such as the clotting cascade and the endocrine system, and would apply to many, many other systems. There is simply no evolutionary explanation of any merit to explain how all these irreducibly complex systems could have occurred naturalistically. So I ask: which one of these arguments is most convincing to you, naturalism or intelligent design?

Before you answer that, though, let's examine a few more things. I would ask you to consider this question: What good thing emanates from naturalism? I would submit nothing! For the naturalist, there is no creator and thus no divine authority, which further means there is no higher standard to determine right and wrong; man becomes his own standard and morality becomes situational. In essence, the atheist cannot claim it is wrong to murder, steal, rape, or molest children, or a multitude of other things; other than that it is against the greater good of society, but this too is superfluous, for who determines what is the greater good of society? It then becomes hard to condemn past atrocities committed by such individuals as Hitler, Stalin, or Mao Tse-Tung, and those perpetuated more currently by organizations such as ISIS or Al-Qaeda, for if there is no God, nothing really matters, does it? Even so, atheists themselves will use such descriptions as bad, good, right, wrong, moral, and immoral.

The fact is, there is good and bad in the world and this actually argues for a God. Craig Thomas has stated, "There is a moral nature inherent in man and there is a moral order to the universe." A.E. Taylor has said in *Does God Exist?* that man may draw the "line between right and wrong in a different place, but at least they all agree there is such a line to be drawn," and this could be at least in part what Paul is saying in Romans 2:12-16. Where do you suppose the "moral nature" in man then comes from? Admittedly, our level of morality in today's society seems to be shifting but the point remains, man would have no concept of right or wrong without God.

Think about this. Can materialism really explain the good things that man does for his fellow man? Certainly, there is evil and selfishness all around us and one only needs to watch the nightly news to be reminded of that fact, but there is also a lot of good being done. Our society, for example, tends to protect the most vulnerable citizens and for that matter even our animals, although our descent into materialism is eating away at this too. We have charitable hospitals, food for the hungry, aid for the oppressed, care for the needy, all these things and much more that come from an altruistic impulse in man not explained by materialism. And our country, the United States, always comes to the aid of other countries when disasters occur, even when these countries may be our enemies. I ask again, where does our "need" to help others come from? Certainly not from naturalism or humanism. I see this every day and I'm sure you do too, in random acts of kindness people demonstrate to each other, from such simple things as holding a door open for someone or paying for someone's meal at a restaurant. I'm sure you have been the recipient of some of these acts of kindness and I myself will never forget kindnesses that have been shown towards me by many. So contrary to what Friedrich Nietzsche, the German philosopher and atheist, said in the past, that "Morality is a hindrance to the development of new and better customs: it makes us stupid," it's just not true; and thankfully so, for where would this world be if it was?

We have seen that the naturalists believe science is hindered by theism and claim it can kill scientific investigation, but this is also silly and a fallacious argument. Richard Dawkins has insisted on "conceiving God as an explanatory alternative to science"; an idea that is nowhere to be found in theologic reflection at any depth, says John Lennox. Lennox likens good science to discovering the inner workings of a car. Knowing the engine of a car was designed by Henry Ford would not stop one from investigating scientifically how the engine worked, and in fact might spur one on to do so. He further points out that there is a great difference between "God who is the creator and a god who is the universe." Most of the scientists of the past, as we have noted, such men as Newton, Einstein, Kelvin, Keppler, Maxwell, Pasteur, Boyle, Faraday, and even Werner Heisenberg, amongst many, many more, were all believers in God and yet practiced sound science and discovered many of the principles we hold true today.

Of course, again, as we have seen, the atheist will assert that the cosmologic argument ends in an infinite regress, because to believe in a god requires us to explain who created God, and then explain who created the one who created God, etc., etc. The problem is, as we have noted, those making this argument are materialistic and God is not. The law of cause and effect does not apply to God. God is a spirit with no corporeal body, and has always been and will always be. Psalms 90:2 reads, "Before the mountains were brought forth, or ever you had formed the earth and the world, even from everlasting to everlasting, you are God." Hebrews 3:4 further tells us, "For every house is built by someone, but he who built all things is God." There is a cause for the universe and this should be self-evident. As we have noted, the Kalam argument says nothing comes from nothing and a beginning demands a cause. The universe existing forever violates the second law of thermodynamics, entropy; and the universe being created from nothing violates the first law of thermodynamics, that is, energy can neither be created nor destroyed, only converted into other forms. Only a God who transcends time and space could supersede the laws He Himself created and bring our universe into existence. The logic of cause and effect is the "ultimate canon" of science and without it all science would fail; for what is science; if it is not about finding causes!

The universe is then the grand effect and God the grand cause. Craig Thomas in his paper, *Does God Exist?*, quotes Robert Jastrow, founder and

former director of the Goddard Institute of Space Studies, "The universe
and everything that happened in it since the beginning of time are a grand
effect without a known cause. An effect without a cause? That is not the
world of science; it is a world of witchcraft, of wild events, and the whims
of demons, a medieval world that science has tried to banish. As scientists,
what are we to make of this picture? I do not know."

Well, I think I know. The cause is God. Psalms 19:1-4 states, "The heavens
are telling of the glory of God; and their expanse is declaring the work of his
hands. Day to day pours forth speech. The night to night reveals knowledge.
There is no speech, nor are there words; their voice is not heard. Their line
has gone out throughout all the earth and their utterances to the end of the
world. In them He has placed a tent for the sun." Werner Heisenberg, one
of the key pioneers of quantum physics, famed for the great Heisenberg
principle of uncertainty, and winner of the Nobel Prize in Physics has said,
"The first gulp from the glass of natural sciences will turn you into an atheist,
but at the bottom of the glass God is waiting for you." Perhaps this is why
so many scientists who were former atheists have turned to God.

Finally, we have seen the great evidence for the Bible being the way God
has communicated to mankind in the past. The internal and external proof
is overwhelming. What is so extremely sad to me is that for the nonbeliev-
ing, there remains no hope except for the hope of this world, and Paul tells
us, "If our hope, even in Christ, is in this world only, we are of all men to
be most pitied" (1 Cor. 15:17). For me, it is inconceivable that we were
placed here on earth for no reason; this makes no sense whatsoever. After
spending the better part of Ecclesiastes defining the futility of life and how
all is vanity, Solomon sums it up in chapter 12, "The conclusion, when all
is heard is: Fear God and keep his commandments, because this applies to
every person. For God will bring every act into judgment, everything which
is hidden, whether it is good or evil."

It is just inconceivable that this is all there is. There must be more to
our existence than just to live a few brief years and then to cease to exist
forever. The purpose of man on earth is delineated in Romans 8:18 and
1 Peter 1:17. We know that our time is short lived and we have another
dwelling place waiting for us. Life is simply more than just having many
"friends" on Facebook and being followed by a lot of people on Twitter. It
is more than having all the fun one can enjoy, how many things a person
can acquire, whether your favorite football team wins the championship, or
whether you make all your three-foot putts, or gaining massive wealth, or

funding a 401-K and seeing how much you can leave your children. I am convinced that mankind has a much higher purpose that leads to quality life here and ultimately hereafter. How much more pleasant this world would be if people would follow the Scriptures and the golden rule!

Think of a world where there is no greed, no violence, no covetousness, no envy, no strife. Of course, this is impossible to achieve on earth because of sin. Paul says in Romans 3:23 that all have sinned and fall short of the glory of God, and further, in chapter 14, verse 12 that we will all give an account to God. Yet sin has become an archaic word in our society; and it seems no one is really responsible for his actions. Yes, sin is in the world, will always be in the world, and is diametrically opposed to good. God has not created us to be machines or robots but has given us free will, and we do have the choice to decide between good or evil.

As I write these last few paragraphs I am in Las Vegas attending a medical conference for recertification purposes. Walking the streets I see people dressed in yellow and green phosphorescent vests, mostly men, flipping baseball-like cards, pimping for prostitutes. Outside the casinos are scantily clad women in their showgirl feathery costumes advertising shows. On one of the crossover walkways various people are begging for money or ostensibly entertaining for it. One fellow is even a little bolder, as he is playing a ukulele with a sign that reads, "I won't lie. I need money for weed." Inside the casinos are people from various walks of life, young, old, disabled, and many others playing mindless games, hoping by chance they may strike it rich.

Walking down the city of glitz, glamor, greed, and decadence, I see myriads of people, many with adult beverages in hand, from all corners and cultures of the world, and I see street vendors selling everything from light-up t-shirts to hundred dollar drives in a Lamborghini; and I doubt few, if any, of these ever considered intelligent design, irreducible complexity, or the cosmologic argument, yet each does have a world view and all will eventually die. The paradox is, if the naturalist is correct, he will never know if his world view is right; however the theist will, along with everyone else, when they realize the final truth.

This all gives me a sense of pathos. I'm reminded of a song from years ago by Peggy Lee called "Is That All There Is?" I'm not saying life has to be a bore or not enjoyable, because I believe it certainly can be exciting and pleasurable in a good sense, but even so, is that what it is all about? If it is, in the words of the song we should all "go dancing, bring out the booze, and have a ball." No, mankind does have a purpose, and in the end every living soul will know that.

We know from the book of Revelation, that in the end God prevails, and God and His people will be triumphant over sin and Satan. You and only you decide which you will side with. This I know from Romans 5:8, "But God demonstrates his love towards us, in that while we were yet sinners, Christ died for us." The saddest people in the world, as I've stated before, are those without hope. The happiest are just the opposite and are filled with hope and joy. The apostle Paul in 2 Corinthians, 11:23-27 was willing to endure hardship, torture, even to death, not because he wanted to but because he understood the joys that awaited him, and counted it joy to suffer as a Christian. Paul knew there was more to life and his reward transcended the carnal. He told Timothy in his second epistle to him in chapter 4, verses 7 and 8, "I have fought the good fight, I have finished the course, I have kept the faith: In the future there is laid up for me the crown of righteousness which the Lord, the righteous judge, will award to me on that day: and not only to me, but also to all who have loved his appearing."

God's requirements are not difficult to understand and obey. Jesus said in Mark 16:16, "He who has believed and is baptized shall be saved." Peter, in the great sermon of Acts 2:38, told those who were asking what they needed to do to be saved from their sins to "repent and let each one of you be baptized in the name of Jesus Christ for the forgiveness of sins... ." My hope and indeed my prayer is that those who have read this book will look seriously at the arguments made in it and determine which view is more plausible, and apply that to their lives. In closing this book, I think it is fitting to look at the beginning words of John in his great gospel:

> [1] In the beginning was the Word, and the Word was with God, and the Word was God. [2] He was in the beginning with God. [3] All things came into being by Him, and apart from Him nothing came into being that has come into being. [4] In Him was life, and the life was the light of men. [5] And the light shines in the darkness, and the darkness did not comprehend it.

Bibliography

Ashton, John F. *In Six Days: Why Fifty Scientists Choose to Believe in Creation*. Green Forest, AR: Master Books. 2001. Print.

_____. *On the Seventh Day: Forty Scientists and Academics Explain Why They Believe in God*. Green Forest, AR: Master, 2002. Print.

Baugh, Carl E. *Why Do Men Believe Evolution against All Odds?* Oklahoma City, OK: Hearthstone Pub., 1999. Print.

Behe, Michael J. *Darwin's Black Box: The Biochemical Challenge to Evolution*. N.p.: Simon & Schuster, 1996. Print.

_____. *The Edge of Evolution: The Search for the Limits of Darwinism*. New York: Free, 2007. Print.

Behe, Michael J., William A. Dembski, and Stephen C. Meyer. *Science and Evidence for Design in the Universe: Papers Presented at a Conference Sponsored by the Wethersfield Institute, New York City, September 25, 1999*. San Francisco: Ignatius, 2000. Print.

Carroll, Sean. *Mysteries of Modern Physics: Time*. N.p.: Great Courses Teaching, 2012. Print.

Collins, Francis S. *The Language of God: A Scientist Presents Evidence for Belief*. New York: Free, 2006. Print.

Darwin, Charles. *Descent of Man*. United States: Pacific Publishing Studio. 2011.

Darwin, Charles Robert, and Michael T. Ghiselin. *On the Origins of Species: By Means of Natural Selection or the Presevation of Favoured Races in the Struggle for Life*. Mineola: Dover Publications, 2006. Print.

Davidheiser, Bolton. *Evolution and Christian Faith*. Grand Rapids: Baker Book House, 1969. Print.

Davies, Norman. *Europe: A History*. Oxford: Oxford UP, 1996. Print.

Dawkins, Richard. *The Blind Watchmaker: Why the Evidence of Evolution Reveals a Universe without Design*. New York: Norton, 1996. Print.

_____. *The God Delusion*. Boston: Houghton Mifflin, 2006. Print.

Dembski, William A. *The Design Inference: Eliminating Chance through Small Probabilities*. Cambridge: Cambridge UP, 1998. Print.

_____. *Mere Creations: Science, Faith & Intelligent Design*. Inter-Varsity Press. 1998. Print.

Dembski, William A., and Michael Ruse. *Debating Design: From Darwin to DNA*. Cambridge University Press. 2004. Print.

Denton, Michael. *Evolution: A Theory in Crisis*. Bethesda, MD: Adler & Adler, 1986. Print.

Faulkner, Danny R. *Universe by Design: An Explanation of Cosmology and Creation*. Green Forest, AR: Master, 2004. Print.

Feynman, Richard P., Robert B. Leighton, and Matthew L. Sands. *Six Easy Pieces: Essentials of Physics, Explained by Its Most Brilliant Teacher*. Reading, MA: Addison-Wesley, 1995. Print.

Gillen, Alan L. *The Genesis of Germs: The Origin of Diseases & the Coming Plagues*. Green Forest, AR: Master, 2007. Print.

Gish, Duane T. *Evolution, the Fossils Say No!* San Diego, CA: Creation-Life, 1978. Print.

Gonzalez, Guillermo, and Jay Wesley Richards. *The Privileged Planet: How Our Place in the Cosmos Is Designed for Discovery*. Washington, DC: Regnery Pub., 2004. Print.

Ham, Ken. *The New Answers Book* (1, 2, 3, and 4). Green Forest, AR: Master, 2006, 2008, 2010, 2013. Print.

Hawking, Stephen, and Leonard Mlodinow. *A Briefer History of Time*. New York: Bantam, 2005. Print.

Hodges, James. *Creation vs. Evolution*. Florida College, 2014. Print.

Humphreys, D. Russell. *Starlight and Time: Solving the Puzzle of Distant Starlight in a Young Universe*. Green Forest, AR: Master, 1994. Print.

Isaacson, Walter. *Einstein: His Life and Universe*. New York: Simon &

Schuster, 2007. Print.

Johnson, Phillip E. *Darwin on Trail.* Washington, D.C.: Regnery Gateway, 1991. Print.

Kirkwood, Bo. *Unveiling The Da Vinci Code: The Mystery of The Da Vinci Code Revealed, a Christian Perspective.* Arizona: Selah Pub. Group, 2005. Print.

Lennox, John C. *God's Undertaker: Has Science Buried God?* Oxford: Lion, 2007. Print.

Lubenow, Marvin L. *Bones of Contention: A Creationist Assessment of Human Fossils.* Grand Rapids, MI: Baker, 2004. Print.

McDowell, Josh. *The New Evidence that Demands a Verdict.* Nashville, TN: Thomas Nelson, 1999. Print.

Meyer, Stephen C. *Darwin's Doubt: The Explosive Origin of Animal Life and the Case for Intelligent Design.* New York: Harper Collins, 2013. Print.

_____. *Signature in the Cell: DNA and the Evidence for Intelligent Design.* New York: HarperOne, 2009. Print.

Miller, Kenneth R., and Joseph S. Levine. *Biology.* Upper Saddle River, NJ: Prentice Hall, 2000. Print.

Moreland, James Porter. *Christianity and the Nature of Science: A Philosophical Investigation.* Grand Rapids, MI: Baker Book House, 1989. Print.

Morris, Henry M. *The Genesis Record: A Scientific and Devotional Commentary on the Book of Beginnings.* Grand Rapids: Baker Book House, 1976. Print.

Morris, John D. *The Young Earth.* Colorado Springs, CO: Master, 1994. Print.

Parker, Gary, and Mary M. Parker. *The Fossil Book.* Green Forest, AR: Master, 2005. Print.

Petersen, Dennis R. *Unlocking the Mysteries of Creation: The Explorer's Guide to the Awesome Works of God.* El Dorado, CA: Creation Resource Publications, 2002. Print.

Polkinghorne, J. C. *Science and Theology: An Introduction.* London: SPCK, 1998. Print.

Rehwinkel, Alfred M. *The Flood in the Light of the Bible, Geology and Archaeology*. Saint Louis: Concordia Pub. House, 1951. Print.

Ross, Hugh. *The Creator and the Cosmos: How the Greatest Scientific Discoveries of the Century Reveal God*. Colorado Springs, CO: NavPress, 1993. Print.

Sailhamer, John. *Genesis Unbound: A Provocative New Look at the Creation Account*. Sisters, OR: Multnomah, 1996. Print.

Schroeder, Gerald L. *Genesis and the Big Bang: The Discovery of Harmony between Modern Science and the Bible*. New York: Bantam, 1990. Print.

Scott, Eugenie Carol. *Evolution vs. Creationism: An Introduction*. Westport, CT: Greenwood, 2004. Print.

Strobel, Lee. *The Case for a Creator: A Journalist Investigates Scientific Evidence that Points toward God*. Grand Rapids: Zondervan, 2004. Print.

Ward, Peter D., and Donald Brownlee. *Rare Earth: Why Complex Life Is Uncommon in the Universe*. New York: Copernicus, 2000. Print.

Weiner, Jonathan. *Planet Earth*. Toronto: Bantam, 1986. Print.

Wells, Jonathan. *Icons of Evolution: Science or Myth?: Why Much of What We Teach about Evolution Is Wrong*. Washington, DC: Regnery Pub., 2000. Print.

Wolfson, Richard. *Physics in Your Life*. Chantilly, VA: Teaching, 2004. Print.

Other Resources
Egan, Scott P. and Ronald Wetherington. "It's Time for Science Texts to Evolve." *Houston Chronicle (Houston, TX)*. 17 Nov. 2013.

Thomas, Craig. "Does God Exist?" Unpublished paper. 2014.

Van Biema, David. "God vs. Science." *Time*. 5 Nov. 2006.

Videos
Can the Biblical Account of Creation Be Reconciled with Scientific Evidence Today? Ankerberg Theological Research Institute, 2004. DVD.

Einstein's Relativity and the Quantum Revolution: Modern Physics for Non-Scientists. The Teaching Company, 2000. DVD.

Evolution vs God. Creation Science Evangelism, 2008. DVD.

Expelled No Intelligence Allowed. Premise Media Corporation Rampant

Films, 2008. DVD.

Impossible: Physics Beyond the Edge. The Teaching Company, 2010. DVD.

Journey Toward Creation. Reasons to Believe, 2003. DVD.

Mysteries of Modern Physics: Time. The Great Courses. 2012. DVD.

Planet Earth. BBC. 2006. DVD.

The Privileged Planet. Illustra Media, 2010. DVD.

Unlocking the Mystery of Life. Illustra Media, 2002. DVD.

Why Is the Big Bang Evidence That God Created the Universe? Ankerberg
   Theological Research Institute, 2004. DVD.

CPSIA information can be obtained at www.ICGtesting.com
Printed in the USA
LVOW11s0216100616

R10999400001B/R109994PG491875LVX1B/1/P